Turning Down the Heat

Also by Hugh Compston

THE NEW POLITICS OF UNEMPLOYMENT (*edited*)

SOCIAL PARTNERSHIP IN THE EUROPEAN UNION (*edited with Justin Greenwood*)

POLICY CONCERTATION AND SOCIAL PARTNERSHIP IN WESTERN EUROPE (*edited with Stefan Berger*)

HANDBOOK OF PUBLIC POLICY IN EUROPE: Britain, France and Germany (*edited*)

KING TRENDS AND THE FUTURE OF PUBLIC POLICY

Also by Ian Bailey

NEW ENVIRONMENTAL POLICY INSTRUMENTS IN THE EUROPEAN UNION

Turning Down the Heat

The Politics of Climate Policy in Affluent Democracies

Edited by

Hugh Compston

and

Ian Bailey

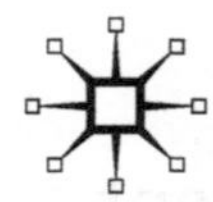

First published 2008 by
PALGRAVE MACMILLAN

Palgrave Macmillan in the UK is an imprint of Macmillan Publishers Limited, registered in England, company number 785998, of Houndmills, Basingstoke, Hampshire RG21 6XS.

Palgrave Macmillan in the US is a division of St Martin's Press LLC, 175 Fifth Avenue, New York, NY 10010.

Palgrave Macmillan is the global academic imprint of the above companies and has companies and representatives throughout the world.

Palgrave® and Macmillan® are registered trademarks in the United States, the United Kingdom, Europe and other countries.

ISBN-13: 978–0–230–20204–7 hardback
ISBN-10: 0–230–20204–7 hardback
ISBN-13: 978–0–230–20205–4 paperback
ISBN-10: 0–230–20205–5 paperback

This book is printed on paper suitable for recycling and made from fully managed and sustained forest sources. Logging, pulping and manufacturing processes are expected to conform to the environmental regulations of the country of origin.

A catalogue record for this book is available from the British Library.

Library of Congress Cataloging-in-Publication Data
Turning down the heat: the politics of climate policy in affluent democracies / edited by Hugh Compston and Ian Bailey.
p. cm.
Includes index.
ISBN-13: 978–0–230–20204–7 (hdbk)
ISBN-13: 978–0–230–20205–4 (pbk)
ISBN-10: 0–230–20204–7 (hdbk)
ISBN-10: 0–230–20205–5 (pbk)
1. Climatic changes—Political aspects. 2. Climatic changes—Environmental aspects. 3. Environmental justice. 4. Greenhouse gases—Environmental aspects. 5. Climatic changes—International cooperation. I. Compston, Hugh, 1955– II. Bailey, Ian, 1965–
QC981.8.C5T87 2009
363.738'74—dc22 2008030135

10 9 8 7 6 5 4 3 2 1
17 16 15 14 13 12 11 10 09 08

Printed and bound in Great Britain by
CPI Antony Rowe, Chippenham and Eastbourne

Contents

Acknowledgements

This book would not have been possible without the time and effort taken by the various contributors to produce high-quality work in a format that enabled the editors to draw comparisons and general conclusions. We would like first and foremost to thank each of the authors for doing this with distinction. We hope that our editorial hand and their hard work have given the volume a level of internal coherence that does justice to the topic while making it accessible to a broad readership.

We would also like to thank others who have contributed ideas that have been used in the book, in particular the participants in two panels on the *Politics of Climate Change* at the ECPR General Conference in Pisa in September 2007, the participants in the *Workshop on the Politics of Climate Change* at the ECPR Joint Sessions in Rennes in April 2008 and, more broadly, the members of the Politics of Climate Change network that links researchers and others across the globe who are interested in exploring strategies for cutting greenhouse gas emissions.

A further debt of gratitude is due to other people who contributed to the production of the book. Particular thanks go to Tim Absalom, Jamie Quinn and Brian Rogers of the Cartographic Resources Unit at the School of Geography, University of Plymouth, for the artwork, and to Amy Lankester-Owen, Gemma d'Arcy Hughes, Geetha Naren and others at Palgrave Macmillan and Integra Software Services for giving us the opportunity to publish this book and for helping throughout the production process.

Finally, because academics are disturbingly prone to taking work home with them, and to staying away from home because of it, we would like to offer a special thanks to our partners and loved ones, Cherrie Summers and the East Trevillis Farmers, Rebecca, Ruth and Polly, for enduring it, we hope, cheerfully.

Notes on the Contributors

Ian Bailey is a Senior Lecturer in Human Geography at the University of Plymouth, UK, specializing in climate governance and the use of economic instruments in UK and EU climate policy. He has published numerous articles on environmental policy in *Annals of the Association of American Geographers, Area, Journal of European Public Policy, Journal of Common Market Studies, Geoforum* and *Environment and Planning C*, and has received research grants from the Economic and Social Research Council, British Academy, and Royal Geographical Society.

Terry Barker is Director of the Cambridge Centre for Climate Change Mitigation Research, Department of Land Economy, University of Cambridge; Leader of the Tyndall Centre's Integrated Modelling programme of research; and Chairman of Cambridge Econometrics. He was coordinating lead author of the Intergovernmental Panel on Climate Change Fourth Assessment Report chapter on the macroeconomic costs of mitigation at national, regional and global levels in the short and medium term. His research interests include greenhouse gas mitigation policy, large-scale computable energy-environment-economy and world energy modelling.

Iosif Botetzagias is a Lecturer in Environmental Politics and Policy at the Department of Environment, University of the Aegean, Greece. His research interests include environmental NGOs and New Social Movements. He recently co-authored a chapter on Greek New Social Movements networking in Purdue D. (ed.), *Civil Societies and Social Movements: Potentials and Problems* (Routledge 2007). He is currently engaged in a transnational project comparing institutions, ideas and interests influencing national policy responses to global climate change.

Paul R. Brewer is an Associate Professor in the Department of Journalism and Mass Communication at the University of Wisconsin-Milwaukee. He is the author of *Value War: Public Opinion and the Politics of Gay Rights* (Rowman and Littlefield 2008). His research on public opinion and political communication has also appeared in such publications as the *American Journal of Political Science*, the *Journal of Politics* and *Public Opinion Quarterly*.

Gary C. Bryner is Professor in the Public Policy Program and the Department of Political Science at Brigham Young University in the United States. He is the author of two books on global environmental issues, *Gaia's Wager* and *From Promises to Performance*, and has also written about US climate policy, carbon trading, market incentives in environmental regulation and sustainable development. He is currently writing a book about how political science theories help to illuminate global environmental problems and policies.

Allison M. Chatrchyan is an Environment Program Leader for the Cornell University extension system and received her PhD in Environmental Politics from the University of Maryland in 2003. In her current position, she participates in the New York State Department of Environmental Conservation/Hudson River Climate Network and TNC Rising Waters Project, and works with Cornell scientists to develop climate change educational materials. She has published widely on legal and environmental policy reform issues in countries with economies in transition.

Hugh Compston is a Reader in Politics at the University of Cardiff, UK, specializing in theoretical and empirical analysis of public policy. He has published four edited volumes on public policy and recently published a single-author study, *King Trends and the Future of Public Policy* (Palgrave Macmillan 2006). He has also published in a number of peer-reviewed journals including the *European Journal of Political Research, Journal of European Public Policy* and *Comparative Politics*, and has received research grants from the Fulbright Foundation, the EU, the Leverhulme Trust and the Economic and Social Research Council.

Chad Damro is Senior Lecturer in European Union Politics and International Relations and Associate Director of the Europa Institute at the University of Edinburgh, UK. His research focuses on the EU's role in competition, trade and environmental policy-making. He has published a number of articles in international peer-reviewed journals on the adoption of greenhouse gas emissions trading in the EU and the EU's shared competency role in international environmental politics.

Deborah Davenport gained her PhD from Emory University (USA) and is Director of the MA Programme in Global Affairs at the University of Buckingham and Lecturer in International Political Economy. She is the author of *Global Environmental Negotiations and US Interests* (Palgrave Macmillan October 2006) as well as numerous other publications on international environmental politics. She has covered climate change negotiations for the *Earth Negotiations Bulletin* since 1997, including the final negotiations of the Kyoto Protocol, and directed the environmental work of the Carter Presidential Center in Atlanta from 1993 to 1997.

Pamela M. Doughman, PhD, is an Energy Specialist with the Renewable Energy Office of the California Energy Commission. Her research focuses on sustainable development and environmental policy. Recent publications include DiMento and Doughman (eds), *Climate Change: What it Means to us, our Children, and our Grandchildren* (MIT Press 2007). Her contributions to this book were made in an individual capacity and not as an employee of the California Energy Commission. The opinions and conclusions expressed do not necessarily express the position, policies or opinions of the California Energy Commission or the State of California.

Lars Friberg is a Research Fellow in international politics at the University of Potsdam, Germany. His research with the Collaborative Research Centre in Berlin focuses on the governance of the Clean Development Mechanism in Brazil and issues related to biofuel certification. Previous to this, he worked with the Climate Action Network Europe. He has been engaged in international climate and sustainable development issues for over a decade.

Irene Lorenzoni is a Lecturer in Environmental Politics and Governance at the School of Environmental Sciences, affiliated with the Tyndall Centre for Climate Change Research, University of East Anglia, UK. Her research focuses on understandings of climate change and energy, with particular interests in how barriers to engaging with climate change influence individual and institutional responses (mitigation and adaptation) and public participation in decision-making.

Douglas Macdonald is a Senior Lecturer at the University of Toronto Centre for Environment. He is the author of two books, *The Politics of Pollution: Why Canadians are Failing their Environment* (McClelland and Stewart 1991) and *Business and Environmental Politics in Canada* (Broadview Press 2007). He has also published a number of peer-reviewed articles and professional works on different aspects of Canadian environmental policy.

Donald MacKenzie works in the sociology of science and technology and markets, especially of financial and carbon markets. He holds a personal chair in sociology at the University of Edinburgh, UK. His most recent books are *An Engine, not a Camera: How Financial Models Shape*

Markets (MIT Press 2006) and *Do Economists Make Markets: On the Performativity of Economics* (Princeton University Press 2007), co-edited with Fabian Muniesa and Lucia Siu.

Sam Maresh was formerly a senior adviser and chief of staff to several ministers in the Victorian and New South Wales state governments. In 2007, he was a visiting scholar at the University of Plymouth, UK, investigating market responses to the EU emissions trading scheme. He has since been contributing to a project investigating governance challenges in response to climate change with the Centre for Energy and Environmental Markets at the University of New South Wales.

Axel Michaelowa teaches international climate policy at the University of Zurich, Switzerland, and is senior founding partner of the consultancy Perspectives. His work focuses on research on, and business in, the Kyoto flexibility mechanisms. He is a member of the Clean Development Mechanism Executive Board and has participated in international climate negotiations since 1995. He was lead author of the Intergovernmental Panel on Climate Change Fourth Assessment Report and is a member of the board of the Swiss Climate Cent Foundation.

Tim O'Riordan is Emeritus Professor of Environmental Sciences at the University of East Anglia in Norwich. He is a member of the UK Sustainable Development Commission, and an advisor to Asda and Anglian Water. He works with the Tyndall Centre to assess the scope for implementing sustainable coasts for East Anglia. He has edited over a dozen books including *Environmental Science for Environmental Management* (Pearson 2000). He is currently looking at the politics of sustainability for a very long way ahead.

Andrew Pease is a Master's Student in mass communication at the University of Wisconsin-Milwaukee. His research focuses on American and Canadian media regulation. He has also conducted research on the effects of celebrity endorsements in United States presidential campaigns.

Nick Pidgeon is Professor of Applied Psychology at Cardiff University and ESRC Professorial Fellow in Climate Change. His research looks at how public attitudes, public trust and institutional responses inform the dynamics of risk controversies. In 2005 he led a major survey on British public perceptions of nuclear power and climate change. He is co-author (with B. Turner) of *Man-Made Disasters*, 2nd ed. 1997, and (with R. Kasperson and P. Slovic) of *The Social Amplification of Risk* (Cambridge 2003).

Joseph Szarka is a Reader in the Department of European Studies at the University of Bath, England. His research interests are in politics and policy studies, where he specializes in environment, climate and energy policy. He is the author of *Wind Power in Europe: Politics, Business and Society* (Palgrave Macmillan 2007) and co-edited *France on the World Stage* (Palgrave Macmillan 2008).

Introduction

Hugh Compston and Ian Bailey

The political problem of climate change

Climate change has emerged as a major scientific and political issue in recent decades, yet progress towards achieving significant reductions in greenhouse gas emissions remains disturbingly slow. Despite the adoption of binding emissions targets by most developed nations, and numerous national climate programmes having been introduced, few governments seem prepared to implement the kinds of policies needed to bring climate change under control (Kerr 2007). The main reasons for this do not appear to be scientific, technological or even economic. Climate science, although still subject to uncertainties, is well established and, via the Intergovernmental Panel on Climate Change (IPCC), has made increasingly clear statements about the scale, impacts and attribution of recent climate change (IPCC 2007a; 2007b). Similarly, economists have developed a range of policy instruments that have the potential to produce major reductions in greenhouse gas emissions without prohibitive economic costs (see Helm 2005; Stern 2007).

The problem, it appears, is essentially political: governments and other political authorities are reluctant to take decisive action even though most are now convinced that strong measures are needed. At present, the main political strategy employed is to develop measures targeting a broad range of emissions sources while, in the main, avoiding or diluting actions that may antagonize business groups or electorates, or which move substantially beyond those taken by other countries. Typical policies thus include setting concrete short-term and aspirational long-term emissions targets, encouraging the development and diffusion of promising technologies, using market mechanisms such as taxes and emissions trading to spur innovation, and urging greater international cooperation on climate policy. Whatever the merits of this approach in terms of maintaining domestic political support and economic competitiveness, the cumulative impact on emissions falls far short of what climate science suggests is needed to avoid serious anthropogenic disruption of the climate system.

One prognosis of the political situation is that increasing scientific and public anxiety about climate change, combined with greater experience in the design of cost-effective policies and instruments, will allow governments to surmount opposition to more ambitious climate policies without a major rethink of political tactics. A more cautious prediction is that the current approach is likely to experience diminishing returns as options for 'no-regrets' emissions reduction become exhausted, and that governments will need to develop innovative political strategies to overcome business threats of disinvestment and electoral resistance to stronger climate measures.

The aim of this book is to contribute towards addressing this issue by identifying political strategies that may enable governments to make major cuts in greenhouse gas emissions without sustaining significant political damage. This is achieved through a broad-ranging analysis of the contemporary politics of climate policy in a range of affluent democracies and at EU level. This book focuses mainly on climate politics and policy at the national and state levels (in the case of federal countries) because this is where the majority of substantive policy measures are formulated and implemented (Bailey 2007), and because the international dimension of climate politics is already well-covered in the international relations and geography literature (see, for example, Grubb *et al.* 1999; Kütting 2000; Vogler 2005; Chasek *et al.* 2006). Affluent democracies provide the main focus of analysis because these are the countries that have contributed most to current greenhouse gas concentrations and which continue, for the present at least, to be the world's largest emitters. They also tend to be the countries where mitigation capacity – in terms of technological and financial resources – and debates on the technical and political dimensions of climate policy are furthest advanced (see Barker, this volume). Although politicians and other key stakeholders directly or indirectly involved in decision-making processes are in many ways the best placed to understand the constraints and opportunities that exist in national climate politics, political scientists also have a contribution to make due to their different and complementary approaches to conceptualizing and analyzing the dynamics of decision-making processes.

Although there is an expanding body of academic writing on climate politics in affluent democracies (see, for example, Helm 2005; Bailey 2007), it does not directly address the question of which political strategies are likely to be of greatest assistance to governments in making deeper cuts in greenhouse gas emissions while minimizing political damage, although a number of strands touch upon it. Accounts of the politics of climate change in various Western countries, for example, often have implications for this question, although these are not always made explicit (see, for example, Bailey and Rupp 2005; Oshitani 2006; Kerr 2007). Similarly, some studies of discourses within environmental politics, and of the political communications strategies used in this area, draw conclusions as to linguistic strategies that are likely to be more effective, but stop short of systematic analysis of political strategies (Hajer and Versteeg 2005; Ereaut and Segnit 2006). The growing literature on climate policies at sub-national level also includes relevant observations (Rabe 2004), while studies of the politics of climate change based on theories of policy-making and political economy have further implications for the development of national political strategies (O'Riordan and Jäger 1996; Newell and Paterson 1998). The extensive literature on international climate politics, meanwhile, often touches upon domestic factors that influence governments' negotiating positions during climate negotiations (Dolzak 2001). Finally, there are studies by social scientists in disciplines such as sociology and psychology which arrive at findings relevant to the analysis of domestic climate politics (Lorenzoni and Pidgeon 2006). However, none of these studies directly address the thorny issue of identifying political strategies that may enable governments to break the apparent impasse between making strong reductions in greenhouse gas emissions and immediate political-economic pressures.

Readership

This book is designed to be of interest to four main groups of reader. First, it will be of direct relevance to policy-makers and activists in affluent democracies (and other countries where climate politics is less developed but beginning to become a major issue) who wish to identify political strategies for developing ambitious climate policies. Second, it provides political scientists and other social scientists with evidence from a number of affluent democracies which can be used to advance the empirical and conceptual analysis of climate policy and politics. Third, it provides a political perspective for natural scientists and economists seeking to overcome political obstacles to the introduction of innovative approaches to manage the main drivers of climate change. Finally, it provides a source of information to postgraduate and undergraduate students in the natural and social sciences who wish to gain a deeper understanding of the political dimension of efforts to control climate change.

Structure of the book

The book is divided into three main sections. The first section sets the scene by reviewing the environmental, technical and economic context of climate politics as it generally affects climate policy in the countries studied in this book, as well as the findings of previous political science studies on the politics of international climate policy. Chapter 1, by Terry Barker, provides the general context of climate policy in affluent democracies and introduces the main policy debates in this area. He first reviews the problem of climate change and its institutional background, then discusses the main economic and environmental impacts of climate change as currently projected and the role of climate policies in relation to climate-change impacts and economic behaviour. This is followed by a review of recent debates on different mitigation strategies and policy instruments; the major generic tensions that have so far dissuaded many governments from taking more ambitious actions to reduce emissions; and the main policy options for mitigation.

Although the existing political science literature on climate policy at the national level does not directly address the question of how political obstacles to more vigorous action on climate change can be overcome, a number of theoretical and empirical studies that examine government responses to other environmental and public policy problems are directly relevant to the climate issue. Chapter 2 by Gary Bryner reviews the political challenges facing governments in relation to climate policy, then discusses a number of perspectives used by political scientists to understand the nature and dynamics of policy-making, focusing in particular on discourse approaches, political economy, policy network theory, policy learning and diffusion, and environmental justice.

Managing climate change effectively will ultimately require international cooperation on an unprecedented scale across a range of policy areas that include energy, transportation, industry, consumption and trade. In particular, international cooperation is essential for addressing the competitiveness fears that have dissuaded many governments from implementing ambitious national programmes. This means that efforts at national (and EU) level to mitigate climate change are necessarily influenced by international developments and that understanding national and EU climate politics requires an appreciation of this global context. Chapter 3, by Deborah Davenport, provides this through a brief narrative history of international developments culminating in the Kyoto Protocol and on to the present day, plus a synopsis of the main findings and debates in the academic literature on the international politics of climate change, focusing especially on work with implications for identifying effective political strategies for reaching international agreements to cut emissions, and the relationship between national and international climate politics.

Chapters 4–14 in Part 2 comprise the core of the book and provide readers with an overview of the politics of climate policy in the EU and the United States, five individual EU member states (the United Kingdom, France, Germany, Sweden and Greece), Australia and Canada, and the United States at state level. Each follows a broadly similar structure. Each chapter begins with a brief narrative history of climate policy in the political jurisdictions concerned that includes consideration of (i) the main policy options advocated or considered but not taken and the reasons for this; (ii) the nature and evolution of technical and economic debates over climate policy; (iii) developments at international level as they relate to the polity being analysed; and (iv) the nature of the political strategies chosen by governments in relation to climate policy. Each chapter then goes on to review the main explanations for these developments, utilizing the academic, expert and political literature as well as the media and focusing in particular on views about the main obstacles to more vigorous action on climate change. Authors then present their diagnosis and analysis of the politics of climate policy in the polity concerned, again focusing on identifying the main obstacles to more ambitious climate policies in different sectors of the economy, and on options to overcome these. Finally, authors summarize their findings for the country in question and draw overall conclusions about what analysis of the polity concerned reveals about the main political obstacles to more effective climate policy and which political strategies are most likely to make it easier for governments to make deep cuts in emissions while avoiding significant political damage.

Across the various countries examined, the authors identify at least six major obstacles to the implementation of more radical climate policies.

The first is the perception that actions by individual countries make little difference to climate change, as this makes it difficult for governments to justify and legitimate the introduction of strong climate policies, especially those that may create additional costs or competitive disadvantages for certain sectors of the economy or society.

The second is the influence of well-financed climate sceptics in undermining support for climate policies by questioning the scientific consensus that climate change is caused by human actions or the economic arguments favouring strong action. Although accumulating scientific evidence appears to be weakening the capacity of climate contrarians to impede climate policy, their influence has been amplified and prolonged in some countries by media framing of climate change as a debate in which the media has a duty to achieve 'balanced' coverage that gives equal exposure to supporters and opponents of the scientific consensus, and by sympathies within some governments for climate sceptic views.

A third obstacle concerns the limitations of many putative solutions to climate change. A number of potential technological fixes, including hydrogen power, nuclear fusion, and carbon capture and storage, are from a technical point of view not yet ready for large-scale deployment or remain uncompetitive under existing pricing arrangements against (often subsidized) fossil-fuel technologies. Other solutions remain contentious in terms of how much they really contribute to removing greenhouse gases from the atmosphere, for example afforestation and carbon capture and storage projects.

Fourth, there are fears about reduced international competitiveness where climate policies impose costs on domestic firms that are not imposed on their foreign competitors. Despite a lack of solid empirical evidence that these fears are justified, industry groups are rarely slow to bring them to the attention of politicians. Such lobbying has often led to the erosion or non-introduction of economic instruments.

A fifth obstacle is the fear of electoral retribution. Growing public sympathy for the general notion of climate protection is arguably a major factor behind the emergence of greater cross-party agreement on the need for stronger climate policies in many countries, but individuals tend to be less supportive of climate policies that directly or indirectly impose personal costs, or which impinge on personal freedoms, such as measures that penalize vehicle use. Democratic governments which ignore these objections risk losing votes at the next election to parties that promise to reverse these policies.

A sixth obstacle concerns divisions in power between different branches of government and the tendency of economic and energy ministries to oppose climate policies that they perceive to have negative economic effects or which threaten established interministerial relations. The ability of such ministries to block or dilute climate policies is enhanced by the fact that responsibility for areas such as energy and transport is generally located in economic, rather than environmental, ministries. While the opposition of economically-oriented ministries is not necessarily enough to block climate policies if heads of government are determined to pursue them, lack of effective leadership at the top is another frequent obstacle to greater progress on climate change.

The authors also identify a number of typical political strategies that are currently being used by governments to strengthen climate policy:

- Efforts to reach global agreements that deepen the commitments made by developed countries to cut emissions and broaden the range of countries making binding commitments, for example by promising financial assistance and technology transfer for developing countries if they agree;
- The use of reporting and target-setting to specify required outcomes, provide statements of intent, build support for action, and inform the structure and design of policy instruments, although reports and targets can be used as a substitute for action, and a recurrent finding emerging from the country chapters is that targets are often missed or eroded;
- A focus on climate policies on which all major relevant actors can agree, as demonstrated by the ubiquity of voluntary agreements whereby industry groups undertake to cut emissions in exchange for the non-imposition of more coercive measures, and the avoidance of policies to which powerful actors, or public opinion, are opposed;
- A preference for incremental changes on many fronts, as opposed to radical changes, in order to avoid arousing political opposition, create a platform for more major changes and allow policy innovations to be tested before being disseminated more broadly;
- Moves to take advantage of spikes in public concern about climate change caused by weather-related natural disasters to introduce or strengthen climate policies without sustaining significant political damage, although these windows of opportunity close quickly once the media agenda moves on;
- Framing climate policy in terms of other policy objectives, for example justifying the expansion of energy generation from renewable sources in terms of improving energy security and increasing employment rather than climate change alone, as this means that actors who support these other objectives can be recruited to swell coalitions favouring, directly or indirectly, the objectives of climate policy.

In terms of policy instruments, in general governments have moved from relying on voluntary agreements towards economic instruments underpinned by legal requirements. Among the most commonly used policy instruments are: information provision; encouragement of new technologies by means of subsidies and grants; encouragement of renewable energy production by means of obliging electricity utilities either to provide a certain proportion of electricity from renewable sources or to buy all electricity produced by renewable sources at a set price (feed-in tariff); regulation to enforce improvements in energy efficiency; emissions trading; and carbon/energy taxes. Important considerations in instrument choice and design have included: effectiveness in reducing emissions in the targeted sector; legal enforceability; ease of monitoring; measures to ensure reasonable equity and, in particular, to protect vulnerable or politically powerful groups; and measures to control the impacts of the policy instrument on international competitiveness.

In the final chapter, the editors draw together the findings of the preceding chapters in order to formulate a number of propositions about the nature of the political strategies that may assist most in hastening progress towards large-scale rapid cuts in greenhouse gas emissions without causing severe political damage to the governments involved. Five main political strategies are identified and discussed.

The first of these is for governments to continue with, and further develop, the political strategies already being used, such as by redoubling efforts to strengthen global agreements in a way that also promotes greater involvement by developing nations; improving reporting of climate change trends and predictions; improving communication of the policy instruments that are needed and the benefits they will produce; introducing progressively stricter emissions and policy targets; identifying and implementing further policies on which all powerful actors can agree; preparing measures that can be implemented swiftly in response to major events that heighten public concern about climate change; continuing to stress the contribution of climate policies to other policy objectives such as energy security; and continuing to incrementally strengthen existing policies, especially economic instruments and financial incentives to promote technological innovations and renewable energy production.

The second strategy is for governments to continue to explore the possibility of introducing new policies such as much more stringent energy-efficiency regulations; much bigger financial incentives for energy-efficiency improvements; *grand projet* style investments in new infrastructure to create step reductions in emissions; extending emissions trading to the individual level by introducing tradable personal carbon allowances; and introducing carbon import tariffs at the EU level to compensate for any losses in international competitiveness caused by the adoption of stringent climate policies.

A third strategy would be to promote reforms in the way that climate policy is governed, in particular by improving the measurement of emissions; devoting more resources to systematic envisioning of what a low carbon society would look like in order to make it easier to identify policies to get there; and improving the integration of economic and environmental governance by means such as moving energy and transport into an environmental ministry. Alternatively, the political profile of climate policy could be raised by creating a separate climate ministry. Other possibilities include providing seats for independent experts and environmental NGOs on all official climate change-related committees on which industry is represented; ensuring that able and committed individuals are placed in key posts; improving the transparency of potentially popular initiatives; and distributing any costs imposed by climate policies more equitably, on the basis that initiatives are more likely to be acceptable if they are perceived as being fair.

Fourth, governments need to identify and implement what we call spillover policies, namely policies that are relatively easily to transfer to other countries, difficult to reverse once introduced, or which create functional or political pressure for their own strengthening or the introduction of related measures.

Finally, although focusing on measures on which the agreement of powerful actors can be obtained has enabled governments to introduce and strengthen some climate policies at relatively low political cost, once the relatively uncontroversial policies have been negotiated and implemented, continued acceptance of the need to obtain broad agreement impedes the introduction of more radical measures by giving stakeholders an effective veto on government action. This, combined with the fact that consensus strategies have not (yet) delivered emissions cuts of the magnitude required to mitigate climate change effectively, suggests that governments which are serious about significantly reducing emissions will at some point need to impose more radical policies against the wishes of powerful actors and/or voters – that is, to adopt a strategy of selective policy imposition. While this approach clearly carries greater political risks than consensus strategies, a number of tactics can be employed to limit these risks. First, introducing unpopular policies during the early years of an administration allows greater time for

Grubb, M., C. Vrolijk and D. Brack (1999), The Kyoto Protocol: A Guide and Assessment, London: The Royal Institute of International Affairs.
Hajer, M. and W. Versteeg (2005), 'A decade of discourse analysis of environmental politics: achievements, challenges, perspectives', Journal of Environmental Policy and Planning 7: 175–84.
Helm, D. (ed.) (2005), Climate Change Policy, Oxford: Oxford University Press.
IPCC (Intergovernmental Panel on Climate Change) (2007a), 'Summary for policymakers', in Solomon, S., D. Qin, M. Manning, Z. Chen, M. Marquis, K. Averyt, M. Tignor and H. Miller (eds) Climate Change 2007: The Physical Science Basis. Contribution of Working Group I to the Fourth Assessment Report of the Intergovernmental Panel on Climate Change, Cambridge: Cambridge University Press, pp. 1–18.
IPCC (2007b), 'Summary for policymakers', in Parry, M., O. Canziani, J. Palutikof, P. van der Linden and C. Hanson (eds) Climate Change 2007: Impacts, Adaptation and Vulnerability. Contribution of Working Group II to the Fourth Assessment Report of the Intergovernmental Panel on Climate Change, Cambridge: Cambridge University Press, pp. 7–22.
Kerr, A. (2007), 'Serendipity is not a strategy: the impact of national climate programmes on greenhouse-gas emissions', Area 39: 418–30.
Kütting, G. (2000), Environment, Society and International Relations: Towards More Effective International Environmental Agreements, New York: Routledge.
Lorenzoni, I. and N. Pidgeon (2006), 'Public views on climate change: European and USA perspectives', Climatic Change 77: 73–95.
Newell, P. and M. Paterson (1998), 'A climate for business: global warming, the state and capital', Review of International Political Economy 5: 679–703.
O'Riordan, T. and J. Jäger (1996), Politics of Climate Change: A European Perspective, London: Routledge.
Oshitani, S. (2006), Global Warming Policy in Japan and Britain: Interactions Between Institutions and Issue Characteristics, Manchester: Manchester University Press.
Rabe, B. (2004), Statehouse and Greenhouse: The Emerging Politics of American Climate Change Policy, Washington DC: Brookings.
Stern, N. (2007), The Economics of Climate Change: The Stern Review, Cambridge: Cambridge University Press.
Vogler, J. (2005), 'The European contribution to global environmental governance', International Affairs 81: 835–50.

Part I
Context

1
Climate Policy: Issues and Opportunities

Terry Barker

Introduction

The purpose of this chapter is to introduce readers to the main environmental, technical and economic debates influencing the politics of climate policy in the countries discussed in this book. The chapter begins by discussing the problem of climate change and its institutional background. This is followed by an assessment of the main economic and environmental impacts of climate change as they are currently projected, so as to establish the political stakes involved and the case for detailed political analysis of national climate strategies. The chapter continues by describing the role of climate policies in relation to climate change impacts and economic behaviour, then outlines recent debates on different mitigation strategies and policy instruments and the major generic tensions that have so far dissuaded many governments from more ambitious actions to reduce greenhouse gas emissions. The main policy options for mitigation that are likely to be the focus of climate policy debates in affluent democracies are then presented, followed by brief conclusions.

Perceptions of the climate change problem

Wide recognition of climate change as a substantial issue in the late 1980s led to the setting up of the United Nations' Intergovernmental Panel on Climate Change (IPCC) in 1988 by the World Meteorological Organization and the United Nations Environment Programme. To quote from the official website: 'The role of the IPCC is to assess on a comprehensive, objective, open and transparent basis the scientific, technical and socio-economic information relevant to understanding the scientific basis of risk of human-induced climate change, its potential impacts and options for adaptation and mitigation' (IPCC 2008).

The IPCC's four Assessment Reports – produced in 1990, 1995, 2001 and 2007 – provide the up-to-date authoritative (but somewhat conservative) consensus from the peer-reviewed literature on the existence, impacts and mitigation options and costs of climate change, and have provided increasingly strong statements about the attribution of climate change to human activities. The 2007 Report concluded that 'warming of the climate system is unequivocal, as is now evident from observations of increases in global average air and ocean temperatures, widespread melting of snow and ice, and rising global average sea level' (IPCC 2007a: 5). The Report also warns of the increasing and long-term risks of serious climate-related damages to water resources, ecosystems, food, coasts and human health (IPCC 2007b). Its conclusion on the *maximum* mitigation costs for the most stringent stabilization range considered (445–535 ppm greenhouse gas concentrations in carbon dioxide equivalent) was a reduction in global Gross Domestic Product (GDP) growth of 0.12 per cent a year to 2050, without including the environmental co-benefits of mitigation, such as reductions in urban air pollution. Nearly all other studies showed much more modest GDP reductions. The cost of this can also be put in terms of the carbon price, defined as the cost imposed on those who release one tonne of carbon dioxide (tCO_2) into the atmosphere. At present it is estimated that real carbon prices (in 2000 prices) are likely to be about \$100/tCO_2 by 2030, but if climate stabilization at 445–535 ppm greenhouse gas concentrations in carbon dioxide equivalent is to be achieved, they would have to rise thereafter, and it would be safer if the carbon price reached \$100/tCO_2 by 2020 rather than by 2030 (see Barker and Jenkins 2007 for further discussion of this issue). The Report nevertheless makes it clear that well-designed mitigation policies could produce higher GDP growth and development than without such policies (IPCC 2007c: 16).

Despite this scientific consensus, there remains a vociferous and well-funded lobby of climate sceptics, often with links to the Republican Party and extreme right in the USA (Mooney 2005), which has sought to influence the public and politicians on climate policy. Their greatest success came in 1997, when the US Senate voted unanimously to reject the Kyoto Protocol because it would seriously harm the US economy and did not impose targets on developing countries. The George W. Bush Administration has since followed this reasoning, with largely rhetorical policies evolving into non-binding and weak emission-intensity targets and technological agreements that emphasize energy security and economic growth rather than climate issues (Blanchard 2003). This is despite evidence available in 2001, when the Bush Administration formally refused to participate in the Kyoto process, which concluded that 'provided policies are expected, gradual and well designed... the costs for the US of mitigation are likely to be insignificant' (Barker and Ekins 2004: 53). Lasky (2003) also concluded that the costs of the USA joining Kyoto would fall from the four per cent of GDP for 2010 quoted by the administration in 2001 to 0.2 per cent of GDP with global permit trading. The evidence for economic harm cited by the Congress and the Bush Administration thus came from the selective use of results based on extreme assumptions.

The Stern Review, conducted in 2006 and published in book form in 2007, has, however, dramatically changed political thinking on the economics of climate change (Stern 2007). While the IPCC reports are policy-relevant but not policy prescriptive, the Stern Review is emphatically prescriptive, leading some commentators to condemn it as 'political' with 'advocacy as its purpose' (Nordhaus 2007: 140). Stern was appointed by the UK Government to advise on the economics of climate change on the basis of a review of the most recent literature available in 2006. The Review proposed a range of climate stabilization targets to avoid the worst climate impacts and the risks of excessive mitigation costs. Since the costs of inaction (5–20 per cent of global GDP forever) were estimated to be many times the costs of action (−1 to 3.5 per cent by 2050), the Stern Review naturally concluded that the global community should act immediately. This argument, supported by the 2007 IPCC Report, has been fully accepted by most Organization for Economic Cooperation and Development (OECD) governments and provides the justification for ambitious political targets at the global level, such as the G8 target set in June 2007 of cutting emissions by 50 per cent by 2050; at the regional level, such as the EU's unilateral 20 per cent target for 2020, rising to 30 per cent if the USA takes comparable action; and by individual countries, such as the UK's 60 per cent target for 2050.

In summary, the problem is clear and the solutions appear to be almost costless. Decisive action will nevertheless require the long-term transformation of the global energy system and is likely to encounter strong opposition from powerful vested interests in the oil, gas and coal sectors as well as the significant group of countries, including the USA, China, Russia and the OPEC nations, where fossil fuels are a major or dominant source of income. It is significant in the political economy of climate change mitigation that the oil, gas and coal sectors in industrialized countries are among the most capital-intensive sectors in the economy and have the highest profit streams with which to fund investment. Conversely, only a very small proportion of these profits need to be channelled into protecting their interests for these sectors to have political influence far beyond their overall contribution to the economy.

Economic, social and environmental impacts of climate change

The climate change problem

The climate change problem is essentially one of accumulating stocks of greenhouse gases in the atmosphere. As is well known, economic behaviour and the availability of fossil fuels have led to greatly increased greenhouse gas emissions from human activity, the unrestrained increase of which in the future is likely to result in dangerous climate change. The main reason to be pessimistic about future emissions is the very substantial reserves of fossil fuels across the world, especially coal, that can be made available at competitive prices for power generation. This has become even more the case with rising gas prices in recent years, gas being a relatively low-emissions fossil-fuel source compared with coal. Adding to this economic pressure to use coal, there are the political pressures for countries that would otherwise import gas to use domestic coal to maintain or increase energy security. Deforestation is another major contributor to greenhouse gas emissions, although the drivers of this are more complex. The loss of virgin forests and grasslands, which is a very long-term global trend, has taken place partly as a consequence of their availability as common resources, so that forest destruction for land or timber benefits individuals but the loss of forest resources and their associated climate change costs are felt collectively.

Impacts of climate change

The most comprehensive summaries of knowledge about the potential impacts of climate change published to date are again those provided by the Stern Review and the Report of the IPCC's Working Group 2. The first impacts of anthropogenic climate change appear to be already evident in the European heat wave of 2003, the Katrina hurricane of 2005 and the widespread fires in Greece and California in 2007. Although the variability of extreme weather events makes direct attribution to climate change difficult, these events are broadly consistent with there being higher average temperatures and more energy in the atmosphere as a result of higher greenhouse gas concentrations. Attribution of such events to global warming is further supported by the unexpectedly high increase in CO_2 concentrations reported by Raupach *et al.* (2007), which they in turn ascribe to faster-than-expected global economic growth and increased use of coal in China for electricity generation.

The important feature of future climate change is an expected increase in the frequency and severity of such extreme events, while rising average temperatures and sea levels should be seen as indicators of the risk of such events, rather than as widespread small and gradual changes. What may appear to be a favourable outcome, such as a milder Northern European climate, may also result in more variable seasons and increasingly frequent and severe floods and droughts. This outlook suggests the character of the political groups that see themselves as being affected by climate change: householders who live at sea level or in flood plains; young people and those with a concern for, or interest in, future generations (such as pension funds); and sectors of the economy that are in some way weather-affected, such as agriculture, water supply, tourism, transport, insurance and construction.

Policies for adaptation and mitigation

The major problem for climate policy is that the atmosphere is a common resource, whereas the benefits of releasing greenhouse gases into the atmosphere are primarily individual. Stern describes this as the greatest market failure the world has ever known. In the economic behaviour underlying current patterns of growth and development, no government, business or household has a direct interest in reducing emissions, but each has an interest in using the atmosphere to dispose of waste gases from combustion. Action by any single country to reduce this will have a very small effect on the global stock of greenhouse gases and, thus, cooperation with others is imperative to reduce costs and achieve substantial reductions.

Climate policy can be broadly divided into adaptation and mitigation policies. Adaptation will take place by societies in response to the environmental impacts of climate change on human and natural systems, and will be both autonomous, such as households protecting their dwellings from local flooding, and via government initiatives, such as flood defences in areas threatened by higher sea levels. Adaptation actions are thus designed to reduce (but not entirely avoid) some climate change impacts and involve both benefits and costs. Net climate change costs are the adaptation benefits less adaptation costs, plus the costs of the unavoided impacts. Mitigation policies differ from adaptation policies in that they reduce emissions at the start of, and throughout, the cycle of change. This is important because of the many unknowns and uncertainties in the effects and feedbacks of climate change; in consequence, mitigation reduces the risks of dangerous outcomes more than adaptation, and also reduces the level of adaptation required. The primary benefit of mitigation is avoided climate change but there are also costs, such as more expensive energy, and co-benefits, such as reduced air pollution and greater rural employment in biomass projects.

Adaptation policies are also geographically distinct from mitigation policies in that adaptation is a local issue associated with ameliorating local environmental and climate conditions. Vulnerable countries, businesses and households have a direct interest in protecting themselves, and the challenge for policy is to identify and reduce risks together with providing the necessary resources. In contrast, mitigation is a global issue because greenhouse gases diffuse quickly over the atmosphere so that mitigating action anywhere reduces overall concentrations, and the issue for policy is to find options that are effective, efficient and equitable. The timing of policies is also different: adaptation has to continue indefinitely and will escalate as extreme climate events proliferate; mitigation, on the other hand, must take place urgently at a global scale if it is to be cost-effective in reducing emissions and, if necessary, reducing the stock of greenhouse gases already in the atmosphere, for example by carbon capture and storage.

Cooperating to avoid dangerous climate change

The context of climate policies is the evidence that accumulating stocks of greenhouse gases creates a risk of irreversible dangerous climate change. The word 'dangerous' comes from the 1992 UNFCCC, whose stated objective is 'to achieve stabilization of greenhouse gas concentrations in the atmosphere at a low enough level to prevent dangerous anthropogenic interference with the climate system'. For the purposes of developing policy, the objective is interpreted by governments, meeting in the UNFCCC bodies, as a political and social agreement. For climate science, dangerous implies climate conditions not experienced for millions of years happening relatively abruptly in geological time and risking potentially damaging consequences in the form of floods, droughts, storms and sea level rise as ice sheets melt. For social sciences, meanwhile, dangerous implies severely damaging outcomes such as flooding or drought that affect social cohesion, long-term economic development and migration, as well as catastrophic local outcomes for communities like small island states that risk losing their land to the sea.

Effective mitigation action has to be global in the form of emission reductions of at least 50 per cent below 1990 levels by 2050, as agreed by the G8 meeting in Heiligendamm, Germany, in June 2007. Since emissions are expected to grow, if nothing is done the implication is that the energy and land use systems have to be comprehensively transformed to produce global emissions reductions of 80–90 per cent below *projected business-as-usual levels* by 2050. Collective action is thus required on an unprecedented scale. The IPCC 2007 Report makes clear that even these reductions have only a 50 per cent chance of achieving the EU's target of maintaining global mean temperatures within 2 °C above pre-industrial levels. The corollary of this is that the global economy must effectively be completely decarbonized and, to be reasonably cautious, new technologies will be needed to take greenhouse gases out of the atmosphere.

The imperative for early action is also supported by economic arguments (Barker *et al.* 2007). Technological change is induced by rising carbon prices because investment in low carbon technologies is increased and costs fall as the carbon price rises, leading to higher uptake of these technologies. The earlier actions are taken to make future carbon prices reliable, the higher these investments become and the lower the eventual costs. Investment costs are also reduced if low greenhouse gas technologies are introduced at the earliest design stage rather than retrofitted.

A key issue for mitigation policies, nevertheless, is how to get sovereign nations and different social groups to agree to cooperate on the management of open access resources, or, as Ostrom (1990: 29; 1998) puts it, to identify contexts in which: 'a group of principals who are in an interdependent situation can organize and govern themselves to obtain continuing joint benefits when all face temptations to free-ride, shirk, or otherwise act opportunistically'. Ostrom studied a variety of social groups in different cultures and times to derive the following general conditions under which groups can manage open access resources successfully:

- the people involved recognize the mutual benefit in cooperation;
- the group has low discount rates in relation to future events, so takes account of future effects;

- there is substantial mutual trust in others following the agreed rules and behaviours;
- there is a capacity to communicate;
- there is a possibility of entering into legal agreements regarding the resource, and any property rights are respected and secure;
- arrangements are made for monitoring the use and condition of the resource and for enforcing any agreements.

These conditions also apply within countries in the formation of coalitions between social groups interested in climate policy and are explored in further detail in the various country chapters.

Debates on climate change mitigation and policies

Cost–benefit analysis versus risk assessment

Alongside new evidence on the increasing risks of climate change, a major shift in economic thinking about the costs and benefits of climate change has occurred since the Stern Review in 2006. The traditional economic approach to the problem has been cost–benefit analysis (CBA), which was applied to climate change in the 1990s (Cline 1991; 1992), although CBA operates most accurately where costs and timeframes can be calculated accurately. In simplified terms, the costs of climate change are set against the benefits of mitigation and adaptation policies in a way that allows comparison between policy options. Nordhaus' aggregate modelling (1991; 1994; 2007) has been particularly influential in monetizing and computing discount rates for the unknown and potentially catastrophic risks associated with global climate change. The outcome of his CBA is an 'optimal' rise in global temperatures with an eventual commitment to warming (for example over 6 °C) not seen for millions of years (Hansen 2007) and very modest prescriptions for action in the form of a small 'optimal' carbon tax (see Van den Bergh (2004) and Barker (2008) for critiques). The costs of mitigation adopted by Nordhaus and other neoclassical economists have typically been exaggerated by setting aside co-benefits and assuming the optimal working of the global economy at full employment, so that any policy intervention is costly. Such policy messages have had a rhetorical use by interest groups and governments wishing to exaggerate costs. Even so, very few countries have adopted this tax prescription, opting instead for no-regrets energy efficiency policies (OECD 2007a; 2007b; 2007c; 2007d) to avoid potential losses in international competitiveness despite the lack of evidence to support such losses both before and after the introduction of such taxes (Barker *et al.* 2007).

Stern considered CBA among several approaches and argued that the economics of climate change are more appropriately concerned with risk rather than return, and with the development of technologies for mitigation, both of which have been evident since the early 1990s when scientific assessments of climate change began in earnest. This in turn implies that the economic problem is one of achieving political targets and lowest costs compatible with equity and effectiveness, rather than with the political–scientific problem of choosing the targets themselves.

Bjørn Lomborg, meanwhile, has organized the 'Copenhagen consensus', a group of Nobel prize winners in traditional economics, to promote the idea that global problems (for example HIV) other than climate change are more worthy of funding judged by comparative cost–benefit analysis (Lomborg 2007). There are, however, two insuperable problems with such comparisons. First, given the politics of policy-making, governments do not make explicit choices between alternatives since there is no stable relationship between policy objectives such as reducing greenhouse gases, economic growth, and better health or education, whatever the political complexion of the government or the prevailing consensus about sound policy. Lomborg's attempt to elicit such relationships also fails essentially because the answers differ between countries and social groups. Social decision-making is, therefore, better characterized by the achievement of agreement between people and groups with potential conflicts of interest and limited information. Second, the climate change problem is systemic and its outcomes potentially irreversible, so the long-term system in which choices are made is threatened in a way that undermines simple short-term marginal trade-offs between policy options.

Technology policies versus mandatory CO_2 caps

The US Administration has generally proposed arguments favouring technological agreements rather than mandatory cap-and-trade schemes such as the EU Emission Trading Scheme (ETS). It is argued that cap-and-trade will not produce the fundamental new technologies required for a zero- or low-carbon economy. The problem with this argument is that some technological developments, such as carbon capture and storage, require a carbon price to become economic, and their investment prospects depend heavily on there continuing to be a carbon price over their useful life. The most effective policies thus appear to be those that combine the carbon price signal from an ETS with direct incentives to fund low-greenhouse gas innovation and research and development, for instance by auctioning emission permits and using a proportion of the revenues to provide additional technological incentives.

Unilateral action, competitiveness and carbon leakage

The main argument faced by governments against the implementation of unilateral climate policies by one country is that it would lead to loss of international competitiveness and, worse, that it would be ineffective because the relocation of high emitting industries to countries without emissions constraints would increase overall emissions (carbon leakage). Detailed studies, however, conclude that these concerns are exaggerated (Barker *et al.* 2007). Although carbon pricing tends to reduce the price competitiveness of carbon-intensive sectors, this may be offset by exchange rate adjustments or improvements in non-price competitiveness, while the extent of competitiveness impacts will vary with the international exposure of the sector in question (higher-value sectors will tend to have higher international exposure than lower-value industries). However, as later chapters show, actual policies in many countries include special exemptions for polluting sectors after lobbying by vested interests, which weaken the effectiveness or efficiency of actions. Rather than provide exemptions for sectors threatened by mitigation policies, a better approach is to provide explicit time-limited subsidies to support adjustment to low carbon outcomes.

Options for policies to adapt to and mitigate climate change

Why a carbon price is essential

The main reason why technology alone is very unlikely to produce effective climate change mitigation is the so-called 'rebound effect' (Sorrell 2007: 5). This arises where improvements in energy efficiency reduce the cost of a technology, which then prompts higher use of particular services (for example heat or mobility) that energy helps to provide, so that the energy saving from the innovation is offset by increased energy consumption. Although rebound effects will vary widely in size, any technological breakthrough without a carbon price to deter extra carbon use risks higher energy demand overall with only a weak reduction, if any, in emissions, especially at a global level. This indicates that a carbon price signal is needed to provide a pervasive and long-term signal for investment decisions so that low greenhouse gas options are chosen consistently. Importantly, research and development decisions would also be influenced by the expected future carbon price.

In simple terms, the low-cost trajectories towards stabilization explored in the literature involve a strong expectation that carbon prices will rise to high levels, encouraging the design, deployment and installation of low greenhouse gas investments in energy supply and in activities that demand energy for power, comfort, light and transportation, depending on lifetime. That carbon prices will be low in the near term reduces the cost of premature obsolescence, while the expectation that they will be high later encourages research, development and investment in long-lived low-emissions capital and reduces the risk of investment lock-in. The outcome of such trajectories should be a rapid adaptation of the energy system without excessive costs and the optimization of no-regrets technical and institutional (environmental tax reform) opportunities and the potential for induced technological change. If the policy is successful, eventually no sector will need to pay for carbon because emissions will cease. However, the price signal must be credible, announced in advance and should escalate, say from $10 in 2013 to $100/t$CO_2$ in 2020, in 2000 prices, to provide time for adjustment.

Carbon prices are generated by policy through two main market-based instruments: carbon taxes and emission permit schemes. A carbon tax is a highly specific and targeted way of tackling global warming through the adaptation of established fiscal systems: the administrative and compliance costs are low compared with those of many other taxes, tax revenues will tend to grow with incomes, and expected responses to higher prices are such that revenues will continue to rise even as there is substantial erosion of the tax base as emissions decline. The problem is that such taxes are disliked, particularly by energy-intensive industries.

Alternatively, the externality can be managed by creating a market in legally enforceable rights to emit greenhouse gases, such as that created by the EU ETS, and then restricting those rights and auctioning all or part of them. These allowances can be given to the emitters as an incentive to participate, as occurred for Phases One and Two of the EU ETS, which is a crucial advantage over taxes in terms of reducing industry opposition. However, there are several objections to such schemes: they acknowledge rights that may not have existed previously; no compensation is normally provided for those who will suffer damage from future pollution; the schemes are open to abuse by collusion; and transactions costs can be high, especially for small non-business sectors. So far the schemes have been confined to cover large fixed business uses of carbon, predominantly power generation.

Portfolios of economic instruments for mitigation: carbon prices, low-greenhouse gas incentives and regulation

The literature on mitigation is concerned mainly with quantitative greenhouse gas targets, as required by any stabilization target, which has to be absolute in relation to the prospective stocks of greenhouse gases in the atmosphere. However, the system driving the creation of emissions is a market-based one in which discount rates are as high as 33 per cent and prices play a critical role in allocating resources and encouraging technological change. Low-cost policies thus all require the use of market instruments via carbon prices, combined in portfolios with regulation and subsidies targeted at clear market failures, most critically the general market failure in innovation and its specific failure in energy markets to achieve, for instance, swifter market penetration of hybrid vehicles or exploitation of no-regrets options in building design. The market failure in innovation comes about because those investing, even allowing for patents, are unable to capture all the benefits which accrue to those able to copy and exploit the innovation. In consequence insufficient innovation takes place in a market system (Jaffee *et al.* 2005).

Governments have a wide range of instruments at their disposal to achieve their climate policy targets (see Table 1.1) (IPCC WG3 Report, Chapter 13). The WG3 Report focuses mainly on sectoral options for mitigation and provides a rich source of detail on the economic potential for mitigation at different carbon prices in energy, transport, buildings, industry, agriculture, forestry and waste management. The appropriate policy portfolios for greenhouse gas mitigation will, of course, be specific to countries depending on their political systems, available renewable and other energy resources, and the energy efficiency of existing building and equipment stock. Such portfolios will combine policies and measures to produce outcomes that are effective at achieving their main climate objectives, efficient with low costs, or even benefits, as regards effects on GDP, and equitable in that vulnerable groups will be most likely to benefit. Importantly for policies to achieve a wide social consensus, they should also address other potential social benefits, such as improvements in

Table 1.1 Sectoral policies, measures and instruments that are environmentally effective in respective sectors in national cases

Sector	Policies, measures and instruments	Constraints or opportunities
All sectors	Public R&D investment in low-emission technologies; standards; carbon prices	Vested interests, environmental tax reform
Energy supply	Reduction of fossil-fuel subsidies	Resistance by vested interests
	Taxes or carbon charges on fossil fuels	
	Feed-in tariffs for renewable energy technologies; renewable energy obligations	May be appropriate to create markets for low-emissions technologies
	Producer subsidies	
Transport	Mandatory fuel economy, biofuel blending and CO_2 standards for road transport	Partial coverage of vehicle fleet may limit effectiveness
	Taxes on vehicle purchase, registration, use and fuels; road and parking pricing	Effectiveness may drop with higher incomes
	Influence mobility needs through land use regulations, and infrastructure planning	Particularly appropriate for countries that are building up their transportation systems
	Investment in attractive public transport facilities and non-motorized transport	

Table 1.1 (Continued)

Sector	Policies, measures and instruments	Constraints or opportunities
Buildings	Appliance standards and labelling	Periodic revision needed
	Building codes and certification	Enforcement can be difficult
	Demand-side management programmes	Utilities need to profit
	Public sector leadership and procurement	For example energy-efficient products
	Incentives for energy service companies	Access to third party financing
Industry	Provision of benchmark information	May be appropriate to stimulate technology uptake. Stability of national policy important
	Performance standards	
	Subsidies, tax credits	
	Tradable permits	Predictable allocation and prices
	Voluntary agreements	Independent monitoring
Agriculture	Financial incentives and regulations for improved land management; maintaining soil carbon content; efficient use of fertilizers and irrigation	Can reduce vulnerability to climate change
Forestry/Forests	Financial incentives (national and international) to increase forest area and to maintain and manage forests; land use regulation and enforcement	Constraints include lack of investment capital and land tenure issues
Waste	Financial incentives for improved waste and wastewater management with renewable energy incentives or obligations	May stimulate technology diffusion

Source: Adapted from IPCC 2007c, Table SPM7.

air quality, better human health, higher crop productivity, increased comfort from better insulated buildings or reductions in traffic-related pollution.

It is a great advantage if both adaptation and mitigation policies are inherently equitable because the main and central benefit of mitigation is the avoided costs of climate change while adaptation also avoids the effects of climate change. Climate change damages are therefore focused on those who cannot relocate or otherwise protect themselves against climate-related damages, which normally means those on low incomes, especially in developing countries with relatively large agricultural sectors in flood plains or drought-prone regions. However, there are major exceptions: energy use per capita may be particularly high in low-quality dwellings occupied by low-income households, for example. In such cases the portfolio should include measures to improve the energy efficiency of dwellings.

One complement to the use of market-based carbon prices is the use of traditional regulatory command-and-control measures, which involve agencies (such as Pollution Inspectorates) to fix and force energy and greenhouse gas standards. Climate, air quality and energy-security objectives are all served by technology-forcing policies of the sort pioneered in California over the past 15 years (Jänicke and Jacob 2004). The main objection to these has been their potential inefficiency, but they can still be targeted to correct market failures and support investments that are profitable where social as opposed to private costs and discount rates are applied.

The potential for environmental tax reform

A particular benefit of environmental tax reforms that may be important for economies with chronic unemployment or underinvestment problems is the potential use of carbon tax revenues to reduce taxes on employment and investment. Distortions in current tax systems may be so great that large numbers of jobs could be created at no net fiscal cost and little risk to inflation by a reform of the tax system (Patuelli *et al.* 2005). This is feasible because a 60 per cent cut in emissions equates to only 2.3 per cent a year over 40 years, which means that if appropriate price incentives are in place, and especially if they can be anticipated, economies can move gradually and efficiently towards sustainable emissions levels without sacrificing economic welfare. However, this requires the use of efficient instruments, such as carbon taxation, and also social acceptance of long-term radical changes which disadvantage carbon-intensive lifestyles.

Conclusions

Although mitigation can reduce climate change and the need to adapt, inertia in the climate system caused by the longevity of CO_2 (the main greenhouse gas) in the atmosphere and the slow response of the oceans to changing emissions means that adaptation policies also remain necessary. Additionally, there is a risk that the international cooperation required for successful mitigation within the timeframes needed to avoid dangerous anthropogenic climate change will not be achieved.

The Stern Review and the IPCC 2007 Reports, together with the experiences of early mitigation policies in the EU and its member states, have provided sufficient information and analysis for global policies to be developed. Political, media and public reception of these reports suggests that the seriousness and magnitude of the climate problem is widely recognized and that their key messages on mitigation and adaptation have been understood. However, effective action requires urgent policy cooperation on an unprecedented scale and strong lesson drawing from other successful international treaties, most notably the Montreal Protocol on ozone-depleting substances.

From the preceding analysis, the most effective policies appear to be those that combine carbon price signals from environmental tax reform for small mobile sources and emission trading schemes for large fixed sources of greenhouse gases with direct incentives for investment in reducing barriers to action, and low-greenhouse gas innovation and R&D funded from tax revenues and emission permit auctions. Such portfolios of market-based instruments can be made even more effective if complemented by technological forcing via standards, such as a requirement for carbon capture and storage by a specified date on all new coal plants. There is evidence from energy efficiency studies that many sectors, particularly buildings, have substantial opportunities for no-regrets mitigation but require tailored policies to reduce or remove barriers, with the policies often led by higher enforced standards on energy efficiency. The main obstacles to more ambitious climate policies thus appear to be political rather than scientific, technological or economic.

References

Barker, T. (2008), 'The economics of avoiding dangerous climate change', *Climatic Change* 89, 173–94.

Barker, T. and P. Ekins (2004), 'The costs of Kyoto for the US economy', *The Energy Journal* 25(3), 53–71.

Barker, T. and K. Jenkins (2007), 'The costs of avoiding dangerous climate change: estimates derived from a meta-analysis of the literature', Briefing Paper for the United Nations Human Development Report 2007/2.

Barker, T., I. Bashmakov, A. Alharthi, M. Amann, L. Cifuentes, J. Drexhage, M. Duan, O. Edenhofer, B. Flannery, M. Grubb, M. Hoogwijk, F. Ibitoye, C. Jepma, W. Pizer and K. Yamaji (2007), 'Mitigation from a cross-sectoral perspective', in, Metz, B., O. Davidson, P. Bosch, R. Dave and L. Meyer (eds), *Climate Change 2007: Mitigation. Contribution of Working Group III to the Fourth Assessment Report of the Intergovernmental Panel on Climate Change*, Cambridge: Cambridge University Press, pp. 619–90.

Blanchard, O. (2003), 'The Bush Administration's climate proposal: rhetoric and reality?' http://webu2.upmf-grenoble.fr/iepe/textes/OB_BushPol_IFRI2_03.pdf [6 December 2007].

Cline, W. (1991), 'The economics of the greenhouse effect', *Economic Journal* 101, 920–37.

Cline, W. (1992), *The Economics of Global Warming*, Washington: Institute for International Economics.

Hansen, J. (2007), 'Scientific reticence and sea level rise', *Environmental Research Letters* 2, 024002, 1–6.

IPCC (Intergovernmental Panel on Climate Change) (2007a), 'Summary for policymakers', in Solomon, S., D. Qin, M. Manning, Z. Chen, M. Marquis, K. Averyt, M. Tignor and H. Miller (eds), *Climate Change 2007: The Physical Science Basis: Contribution of Working Group I to the Fourth Assessment Report of the Intergovernmental Panel on Climate Change*, Cambridge: Cambridge University Press, pp. 1–18.

IPCC (2007b), 'Summary for policymakers', in Parry, M., O. Canziani, J. Palutikof, P. van der Linden and C. Hanson (eds), *Climate Change 2007: Impacts, Adaptation and Vulnerability. Contribution of Working Group II to the Fourth Assessment Report of the Intergovernmental Panel on Climate Change*, Cambridge: Cambridge University Press, pp. 7–22.

IPCC (2007c), 'Summary for policymakers', in Metz, B., O. Davidson, P. Bosch, R. Dave and L. Meyer (eds), *Climate Change 2007: Mitigation. Contribution of Working Group III to the Fourth Assessment Report of the Intergovernmental Panel on Climate Change*, Cambridge: Cambridge University Press, pp. 1–23.

IPCC (2008), 'About IPCC', http://www.ipcc.ch/about/index.htm (12 May 2008).

Jaffe, A., R. Newell and R. Stavins (2005), 'A tale of two market failures: technology and environmental policy', *Ecological Economics* 54, 164–74.

Jänicke, M. and K. Jacob (2004), 'Lead markets for environmental innovations: a new role for the nation state', *Global Environmental Politics* 4, 29–46.

Lasky, M. (2003), *The Economic Costs of Reducing Emissions of Greenhouse Gases: A Survey of Economic Models*, Technical Paper Series, Washington DC: Congressional Budget Office.

Lomborg, B. (ed.) (2007), *Solutions for the World's Biggest Problems: Costs and Benefits*, Cambridge: Cambridge University Press.
Mooney, G. (2005), *The Republican War on Science*, New York: Basic Books.
Nordhaus, W. (1991), 'To slow or not to slow: the economics of the greenhouse effect', *Economic Journal* 101, 920–37.
Nordhaus, W. (1994), *Managing the Global Commons: The Economics of Climate Change*, Cambridge MA: MIT Press.
Nordhaus, W. (2007), 'The challenge of global warming: economic models and environmental policy', http://nordhaus.econ.yale.edu/dice_mss_072407_all.pdf [7 December 2007].
OECD (Organization for Economic Cooperation and Development) (2007a), *The Political Economy of Environmentally Related Taxes*, Paris: OECD.
OECD (2007b), *Impact of Environmental Policy Instruments on Technological Change*, COM/ENV/EPOC/CTPA/CFA(2006)36/FINAL, Paris: OECD.
OECD (2007c), *Assessing Environmental Policies*, Policy Brief, Paris: OECD.
OECD (2007d), *Instrument Mixes for Environmental Policy*, Paris: OECD.
Ostrom, E. (1990), *Governing the Commons: The Evolution of Institutions for Collective Action*, Cambridge: Cambridge University Press.
Ostrom, E. (1998), 'A behavioural approach to the rational choice theory of collective action', *American Political Science Review* 92, 1–22.
Patuelli, R., P. Nijkamp and E. Pels (2005), 'Environmental tax reform and the double dividend: a meta-analytical performance assessment', *Ecological Economics* 55, 564–83.
Raupach, M., G. Marland, P. Ciais, C. Le Quéré, J. G. Canadell, G. Klepper and C. Field (2007), 'Global and regional drivers of accelerating CO_2 emissions', *Proceedings of the National Academy of Sciences* 104, 10288–93.
Sorrell, S. (2007), *The Rebound Effect: An Assessment of the Evidence for Economy-Wide Energy Savings From Improved Energy Efficiency*, London: UK Energy Research Centre.
Stern, N. (2007), *The Economics of Climate Change: The Stern Review*, Cambridge: Cambridge University Press.
Van den Bergh, J. C. J. M. (2004), 'Optimal climate policy is a utopia: from quantitative to qualitative cost-benefit analysis', *Ecological Economics* 48, 385–93.

2
Political Science Perspectives on Climate Policy

Gary C. Bryner

Introduction

Chapter 1 introduced the main environmental, technical and economic debates affecting the politics of climate policy and highlighted the increasing scientific confidence that the climate is already changing as a result of growing greenhouse gas concentrations. However, despite the near consensus about the basic factors driving climate change, uncertainties about the extent, location and timing of disruptive impacts means that policy-making must take place against a backdrop of uncertainty. Equally, dramatic reductions in emissions – and the changes to energy production and use that this entails – present enormous challenges for politicians, political scientists, economists and other social scientists. Many tools in economics and political science centre on marginal, incremental policy change, but preventing climate disruptions will require fundamental shifts in behaviour and policies. The purpose of this chapter is to suggest frameworks from political science that might facilitate examination of the political changes and policy innovations likely to be required to address climate change. It begins by reviewing the political challenges facing governments in respect of climate policy, then considers a selection of perspectives utilized by political science to understand the nature and dynamics of policy-making, focusing in particular on discourse approaches, political economy, policy network theory, policy learning and diffusion, and environmental justice.

The political challenges of climate change

There are growing signs that the emerging consensus among climate scientists is beginning to be reflected in policy commitments throughout large swathes of the industrialized world: the European Union (EU), along with increasing numbers of national and subnational governments, is setting and tightening emissions targets, while the majority of industrialized nations agreed in the Kyoto Protocol to reduce their collective greenhouse gas emissions to 5.2 per cent below 1990 levels by 2008–2012. As is well known, the USA and Australia subsequently withdrew from the effort (although Australia rejoined in 2007), and the EU is the only major emitter that may achieve its Kyoto target. Emissions in the USA have increased by 16 per cent between 1990 and 2005 (US Environmental Protection Agency 2006), but the failure of most industrialized nations to meet even their modest Kyoto targets makes all the more daunting the political challenge of climate change.

The political transition to more effective climate policies will be particularly difficult because these are likely to require energy prices to climb rapidly even as scientific uncertainties persist and the benefits of some preventative actions lie relatively far in the future. This will be particularly difficult in the USA, where raising taxes of any description is anathema to many politicians and organized interests. Efforts to raise energy taxes have been more acceptable in Europe, where petrol prices remain several times higher than in the USA, although further rounds of energy tax increases in Europe are likely to encounter opposition if the USA continues its recalcitrance towards binding climate commitments and current fears of economic recession prove well-founded.

Either way, climate policies will need to go far beyond higher energy taxes to stand a reasonable chance of constraining mean global temperatures to within 2 °C above pre-industrial levels. Pacala and Socolow (2004) calculate that seven billion tonnes of carbon emissions will need to be avoided by 2050 to achieve a temporary stabilization of greenhouse gas concentrations in the atmosphere before new technologies become available in the second half of the twenty-first century. They then identify options that would each yield one billion tonnes of carbon savings, ranging from doubling the fuel efficiency of light vehicles to reducing the number of miles driven by half and an 80-fold increase in wind power or a 700-fold increase in photovoltaic power. The magnitude of each action underlines the scale of the political challenge of climate change.

Unless major breakthroughs in clean energy technologies or engineering occur, it seems clear that major restructuring of economic life will also be needed. Suburban sprawl, single occupant vehicles, high energy-use recreation, food shipping, and the production and use of many energy-intensive goods and services are likely to have to be dramatically curtailed. Climate change, plus policies such as biofuel production expansion, may also lead to falling food production unless land is diverted from other uses. Meanwhile, governments will still have to contend with other social and economic issues (Romm 2007). If scientists were becoming less convinced about the seriousness of climate change, it would make sense to delay expensive policy responses until the evidence was more certain, but just the opposite is true, so governments inevitably face a game of political risk management in how they approach climate policy.

Discourse approaches to framing climate change

Discourse approaches have long been used by political scientists and scholars in other disciplines to explore the various ways politicians and societies 'define, interpret, and address environmental affairs' (Dryzek 1997: 10). Discourses represent shared ways of comprehending the world: the more complex and contested a situation, the more useful discourse analysis can be in understanding different perspectives and facilitating exchanges of ideas. The particular relevance of discourse analysis here is that it draws attention to the different ways in which the political problems and proposed solutions relating to climate change have been conceptualized.

Dryzek (1997), for example, proposes a two-dimensional scheme to help distinguish the various discourses that have been applied to human–environment relations. The first dimension focuses on whether industrialization remains a tenable ideology or a radical departure from industrial–consumerist discourses is required. At one end of the continuum, environmental problem-solving takes current social, economic and political structures as given and examines how adjustments can be made to solve environmental problems while, at the other, survivalist discourses argue that business as usual will eventually hit ecological limits, necessitating a radical shift from conventional commitments to economic growth. The second dimension classifies responses according to their level of imagination. Prosaic responses again accept industrialism as given and maintain that responses to ecological problems can be radical or incremental but do not require a new society. Imaginative responses, in contrast, replace assumptions and values underlying industrial society with the idea that environmentalism is not an external constraint but an essential opportunity. The changes that result can be radical or incremental as long as they are built on a new way of thinking about the environment.

Table 2.1 Classifying environmental discourses

Dimensions	**Reformist**	**Radical**
Prosaic	Problem solving	Survivalism
Imaginative	Sustainability	Green radicalism

Source: Dryzek 1997: 14.

Dryzek then combines these two dimensions to identify four distinct discourses (Table 2.1).

Environmental problem-solving accepts the political–economic status quo but seeks to adjust it through policies designed to cope with environmental problems. In relation to climate change this implies that current political and economic arrangements do not constitute significant obstacles to stronger action on climate change. Survivalism suggests that radical changes will be required by means of new political and administrative controls placed on industrialization. Thus, industrialization is seen as a major problem and stronger state control of the economy is seen as a necessary strategy if climate change is to be controlled.

Sustainable development, emphasized since the 1980s, seeks to eliminate conflict between economics and environment. This discourse views limits as no longer determinative and maintains that creative solutions can take place within the existing political economy. This in turn implies that while the free market economy may be problematic in relation to climate change, market-based instruments are also a large part of the solution. Green radicalism rejects industrial society in favour of one based around the primary prerequisite of a healthy environment. However, the debates between deep ecologists, social justice ecologists and other greens over how to reconstitute society to cope with problems such as climate change are lively and diverse (Dryzek 1997).

Each discourse includes alternatives for framing choices. Under environmental problem-solving, for example, problems can be turned over to experts to solve through administrative structures and procedures that 'rationally' assess problems, evaluate options, and design and implement policies. An alternative is to stimulate public debate, so that democratic processes – rather than expert analysis – form the basis of action. Finally, market-based approaches emphasize the importance of getting prices right, recognizing private property rights and promoting competition for solving environmental problems.

Clapp and Dauvergne (2005) refine Dryzek's approach in their fourfold typology of discourses for framing environmental choices. Market liberals champion economic growth and high personal incomes as key to environmental protection: the more wealth that is available, the more can be invested in environmental protection and ecologically sustainable practices. Its advocates recognize that this also contributes to economic inequality, but argue that in the long run all will be better off. They are also generally optimistic about the possibility of technological fixes to most problems and extrapolate from past progress to predict that human ingenuity encouraged by market incentives will deliver improved environmental performance. Institutionalists generally share market liberals' optimism regarding the sustainability of economic growth but also argue for strong political institutions to shape markets, transfer resources, create norms and commitments to protect collective interests, and promote global cooperation (Vogler 2005; Ward 2006). Such institutions, they argue, support the development of shared values and international laws and agreements that become global regimes for governing specific environmental (or other policy) issues (Young 1997).

Bioenvironmentalists, by contrast, reject faith in markets and institutions, arguing that economic and political models of development need to be based around new ways of measuring wealth and economic activity that recognize ecological limits. For many bioenvironmentalists, global government with strong coercive authority is required to shape economic activity (Clapp and Dauvergne 2005). Finally, social greens believe that social inequality, discrimination and domination are intertwined with environmental decline. According to this perspective, economic growth, industrialism, uneven trade relations, overconsumption and poverty combine to threaten ecological sustainability. Their prescription for this is the dismantling of global trade and a return to local autonomy and community-level decision-making that promotes justice and equality and ensures the survivability of indigenous and other marginalized people (Clapp and Dauvergne 2005).

These framings help to focus attention on the key issue of whether climate change can be effectively addressed through some combination of markets and public policies, or whether new forms of social, political and economic organization are needed. Although it may be the case that climate change does require a radical transformational, this carries high political risks because valuable time may be lost in the endeavour and strong opposition is likely from key interest groups and electorates. Current solutions may be piecemeal, incremental and ultimately insufficient, but early action to reduce the most environmentally destructive practices may better position governments and societies to understand whether sufficient innovation can occur within the existing industrialist/market-based paradigm (Bryner 1999).

A political economy of climate change: making markets work

Even if one accepts reformist rather than radical responses to climate change, building support for policy changes will require a transformation in understandings of the political economy of markets. Borman and Kellert (1991: xii) argue that poorly functioning markets have created a global environmental deficit because 'the longer-term ecological, social, and economic costs to human welfare [of appropriate environmental stewardship] are greater than the shorter-term benefits flowing from these alterations'. Such environmental deficits also steal from future generations by permitting profligate consumption by the current generation which disregards the needs of future generations. Other scholars highlight the fact that politics is often plagued by political calculations and pressures which drive incentives to insulate and protect powerful industries rather than forcing them to compete and become economically (and environmentally) efficient (Yandle 1999; Ciocirlan and Yandle 2003). Markets, by contrast, are portrayed by public-choice theorists as provoking innovation, cost effectiveness and expanded choices.

Characterizing politics and markets as polar opposites is an attractive strategy for those wishing to reduce the scope of political decision-making but fundamentally misstates the fact that many markets fail to provide environmental stewardship because powerful interests can externalize costs on third parties that are powerless or too widely dispersed to protect themselves against them. A more constructive approach looks at the intersections between politics and markets. To ensure that their benefits are realized, markets require strong and capable institutions to assign property rights, monitor emissions and enforce requirements. If these conditions can be created, markets can play a major role in reducing the threat of climate change alongside regulations, subsidies, research, education and other policies. Public policies are also required to deal with the distributional consequences of markets, including policies that facilitate adaptation to the impacts of climate change.

Four approaches used by political and other social scientists have proved to be particularly helpful in describing and analyzing the challenges that politicians face in relation to the reformulation of the relationship between markets and the environment: policy network theory; theories relating to policy learning and diffusion; theories that explore policy integration; and theories of environmental justice.

Policy network theory

Policy network analysis and the advocacy coalition framework have emerged in recent decades as significant additions to the toolbox of techniques used by political scientists to interpret the various ways that discourses and power relations permeate and inform environmental decision-making (Jenkins-Smith 1990; Hajer 1995; Sabatier 1998; Smith 2000). Policy network approaches are particularly useful tools for policy analysis because they focus on the meso- and micro levels of decision-making, where a substantial proportion of debates that lead to policy change take place (Zito 2000), rather than the macro level, studies of which reveal relatively little about the actor interactions that shape final decisions.

Policy network and advocacy coalition analyses thus assist in identifying the key governmental and non-governmental actors involved in policy decisions and in structuring understandings of the divisions of power and influence over decision-making (Zito 2000). For instance, government ministries and parliaments exert power through their conferred authority to create policy, whereas industry groups and scientific communities do not possess decision-making power but can wield significant influence through threats of business disinvestment, strategic cooperation to gain competitive advantages or trade-offs on another issue, or authoritative claims to hold policy-relevant knowledge (Haas 1990; Lévêque 1996). So too, in a more diffuse way, public opinion and electoral trends can act as major brakes on, or spurs for, policy action. The approach is also helpful in understanding how discourses gain credibility and disseminate when harnessed by particular actor groups and how discourses in turn shape actor viewpoints. Smith (2000), for example, explores how network actors draw upon and manipulate discourses as a resource in their policy activities while utilizing their power or influence to promote certain ideas (such as sustainable development or ecological modernization), but in so doing themselves develop a tendency to view problems and solutions through the lenses of these same discourses.

Applying policy network analysis to climate policy can therefore provide useful insights into the roles of, and relations between, advocacy coalitions, the concerns they seek to promote, and the resources they are prepared to exchange to achieve desired objectives. Smith (2000) in particular emphasizes the importance of resource interdependencies, along with exogenous forces, to the behaviour of policy networks. A clear example of this approach is Zito's (2000) analysis of the difficulties experienced by the EU in gaining concurrent multi-institutional support for a harmonized EU carbon tax where multiple political and economic objectives are at stake, while Damro and Luaces Méndes (2003) and Wettestad (2005) demonstrate how a coalescence of powerful advocacy coalitions and discourses about the political, environmental and economic desirability of emissions trading made possible the EU's rapid adoption of a policy instrument that it had opposed only a few years before in the Kyoto negotiations.

Policy learning and diffusion

Climate policy is clearly an area where high levels of policy learning and diffusion are required to counteract, among other things, poor understanding of the climate problem, the influence of powerful fossil-fuel interests and the tendency for political ideologies to privilege certain solutions over others (Victor 2004). According to Jordan (2005: 308), policy learning 'involves a cognitive and reflective process in which policy makers adapt their beliefs and positions in view of past experiences (lesson drawing), experiences of others (diffusion), new information and technological developments which actors apply to their subsequent choices of policy goals or techniques'. These concepts are important for three reasons: (i) we cannot assume that the best policies have already been found for most economic and social problems (knowledge problem); (ii) the dynamic development of the world through technological progress and social change implies the perennial emergence of new problems as well as qualitative and quantitative changes in old problems; and (iii) complex political, economic, social and cultural conditions influence diffusion of 'new' ideas from one policy issue or political jurisdiction to another (Fiorino 2001; Weyland 2005; Jacoby 2006). A better understanding of policy learning and diffusion would obviously be beneficial for those wishing to put across new ideas about controlling climate change.

In his classic work on the subject, Hall (1993) argued that policy learning and change occurs at three levels: (i) alterations to existing policy instruments (first-order change); (ii) the adoption of new instruments (second-order change); and (iii) strategic shifts in perceptions of the policy problem and policy goals, often encompassing broader shifts in social attitudes (third-order change). Hall concluded that third-order changes are best understood as paradigm shifts, as they extend change beyond ordinary policy-making, in extreme cases to revolutionize the basis and practice of public policy (see also Carter 2004). Many would argue that this is precisely what is needed in relation to climate change. Hall contends that such changes occur when an existing policy paradigm ceases to provide adequate solutions to a key problem and attempts to remedy this by adjusting existing instruments or by deploying new instruments fail, as mounting evidence of failure may trigger a political contest between competing solutions, followed by the institutionalization of victorious ideas as a new paradigm.

Oliver and Pemberton (2004) argue, however, that Hall's description fails to capture the capacity of old paradigms to defend and adapt themselves, making paradigm change a much more iterative and uncertain process than Hall's typology allowed. New ideas, for instance, may be partially integrated as part of a punctuated evolution of old paradigms. Although Oliver and Pemberton concede that exogenous shocks may trigger paradigm shifts, they maintain that policy and social change is much less predictable than Hall implied as a result of the complex interplay of institutions, actors, interests, policy legacies and policy styles. Similarly, Jordan *et al.*'s (2003) exploration of the diffusion of market-based and voluntary environmental policy instruments between the 1980s and the early twenty-first century charts' highly uneven patterns of instrument adoption and the problems they were deployed to deal with, which they explain in terms of pre-existing national and supranational dependencies.

These conclusions resonate with political science writings from a new institutionalist perspective, which emphasize the importance of policy settings and national governance traditions to governments' willingness to experiment with new policies and political strategies (Richardson and Watts 1986; Linder and Peters 1989; Scott 1995). Wurzel (2002) argues that governments and civil services invest considerable time and resources in developing standard problem-solving procedures and are reluctant to depart from these unless they are clearly dysfunctional. Over time these accrete into distinctive national policy styles, which in turn lead to a tendency for incremental and bounded innovation. Thus, energy taxes aimed at raising revenue, for example, are more acceptable in relatively high-tax countries and political cultures that assume a higher level of government regulation and a larger public sphere of decision-making than is the case in the USA (Vig and Faure 2004).

At the same time, Jordan *et al.* (2005) recognize the weakness of the idea of stable policy styles in decision-making processes whose basic dynamics involve unequal power relations and competing ideas and interests. Although preferred styles may exist, their durability depends on the government in office, the government department leading policy development, which interest groups promote their views most effectively (Howlett and Lindquist 2004), and the issue in question. Both views are well documented by Jasanoff (2005), who highlights the influence of policy style in shaping US and EU policies towards biotechnology while also using the example of how the UK innovated boldly and swiftly to cope with Bovine Spongiform Encephalopathy to show how policy makers confronted with novel and high-profile problems can suddenly move to rapid and strong policy innovation. Jordan *et al.* (2003) also conclude in relation to the deployment of market-based instruments that although states have not responded in precisely the same way to common problems, this may simply be attributable to 'time-lag' effects caused by greater or lesser degrees of compatibility between 'new' instruments and pre-existing national policy styles, and that a general pattern of converging responses still exists.

In addition to the general concept of policy style, scholars have produced numerous studies of other factors that influence policy learning. In particular, studies of how science and technologies inform and interact reciprocally with political processes and institutions illuminate the difficulties with, and possibilities for, policy learning (Jasanoff 1998). These are obviously relevant given the key role of science and technology in climate policy. Other factors affecting learning are deeply rooted in the nature of democratic politics and its ability to address public concerns and needs while, at the same time, engaging with complex scientific and technological issues (Fischer 2003).

Policy integration

The complexity and connectedness of climate change with other issues (energy, transport, waste and so on) means that effective action requires strong policy integration. Dryzek (1997) argues that the traditional response of governments to complex environmental problems is to disaggregate them into their constituent parts and assign policy responsibilities for each component to specialist expert agencies. However, the efficacy of this approach (which he terms administrative rationalism) is contingent on, among other things, weak interactions between problem subsets (the problem's divisibility), and the capacity of decision-making systems to reassemble problem components into a coherent overall solution. Failure to achieve either of these conditions is likely to result in incomplete or disjointed solutions to problems or the displacement of the problem from one medium to another (for example, the promotion of hydrogen fuel-cell technologies that merely shift emissions from vehicle exhaust pipes to power stations rather than reducing them) instead of genuine problem solving.

Empirical studies of environmental policy integration reveal important differences between the USA and the EU. Whereas US environmental policy and law is generally characterized by fragmentation and lack of coordination (Davies and Mazurek 1998) and the idea of sustainable development, which is explicitly aimed at integrating environmental, ecological and equity concerns, has found little support in the USA (Bryner 2000), the EU has strongly embraced the principle of environmental policy integration. As far back as 1973, the first European Community Environmental Action Plan argued for assessment of the environmental impacts of any measure that is adopted or contemplated at national or Community level (Lafferty and Hovden 2003). The 1992 Maastricht Treaty created the further requirement for environmental considerations to be integrated into all other policies, and the principle was assigned its own article in the Amsterdam Treaty in 1997. A critical element of this article is that it gives environmental objectives 'principled priority' over other policy goals. The burden of proof is thus taken 'off the shoulders of those promoting environmental protection... thereby shifting the balance of responsibility to other interests, actors, and objectives' (Lafferty and Hovden 2003: 11).

In order to achieve genuine environmental policy integration, both vertical and horizontal integration must be considered (Lenschow 2002). Vertical integration focuses on the extent to which departments or ministries consider environmental goals as being central to everything within their areas of responsibility from policy creation to implementation. This does not necessarily give primacy to environmental goals (other policy concerns still apply), but it does compel officials to identify, assess and report on all major environmental issues relevant to their portfolios (Lafferty and Hovden 2003). Horizontal integration refers to the cross-sectoral strategy of integrating environmental concerns into the work of every ministry, including the creation of a centralized authority to supervise and coordinate environmental priorities, articulate clear targets and timetables for environmental policy, and ensure the production of regular and thorough environmental impact and strategic environmental assessments that can be used to guide major decisions.

Studies of environmental policy integration in the EU nevertheless reveal the difficulties of developing genuine horizontal and vertical policy integration. In particular, demands for national and regional autonomy within the EU have led to unhelpful policy cleavages, while institutional arrangements that gave unequal powers to sectors which historically took little account of their environmental impacts (such as agriculture, energy and transport) have helped to maintain policy fragmentation (Lenschow 2002). Climate policy integration is thus a problem that governments must address to ensure that activities in one branch or level of government to reduce emissions do not undermine other policy goals or climate initiatives developed in other branches or levels.

Environmental justice

The concept of environmental justice has long been debated by political science and other social sciences as a basis for understanding and determining policies relating to the distribution of environmental risks throughout society (Dryzek 1997; Schlosberg 1999; Adger *et al.* 2006). It focuses on the ways in which certain communities, typically low-income residents and ethnic/racial minority groups, disproportionately bear the environmental costs of industrial activities and, increasingly, climate change. A path-breaking 1987 study, updated in 1994, for example, found that, in the USA, race was the most significant variable in explaining the location of commercial hazardous waste facilities (Charles 1987; Goldman and Fitton 1994).

Climate change raises particularly challenging ethical issues about distributive justice because of the mismatch between those who enjoy the benefits of emissions and those who bear the burdens but lack the resources to adapt to or evade them. Whether it is Pacific island states facing inundation from rising sea levels (Barnett and Adger 2003) or Bangladeshi residents facing flooding and loss of clean water, those consuming the fewest resources and producing the lowest levels of greenhouse gas emissions generally stand to suffer most from disruptive climate change. Environmental justice is also directly relevant to the design of national climate policies, for example, where regressive green tax policies exert a proportionately higher burden on those on lower incomes (Ekins and Barker 2001). Intergenerational equity also becomes an important consideration for long-term problems like climate change.

In addition, pre-existing inequities often combine to exacerbate the negative impacts of climate change. Short-term economic pressures, for example, may cause those experiencing poverty to engage in practices that produce immediate but unsustainable benefits (Adger *et al.* 2006; Hossay 2006). The redistribution of resources, whether by compensation or by concession packages, thus becomes an important consideration in ensuring that climate policies remain reasonably equitable and secure popular support.

Conclusion

While the uncertainties surrounding climate science will continue to be studied and debated, just as important are the debates concerning the kinds of policies required to mitigate and adapt to disruptive climate change, and the political strategies needed to deliver these policy changes. There is little prospect that politics as usual will produce effective solutions. Long-term commitments by many affluent democracies to ambitious greenhouse gas reductions represent a significant breakthrough. However, while such commitments are essential to guide decisions on power plants, buildings and other infrastructure that will have a lifetime spanning decades, these must be accompanied by short-term milestones and the creation of institutions and policy structures capable of responding to the numerous challenges created by climate change. This makes all the more important and pressing the need to understand better the politics required to produce ecologically sustainable climate policies.

References

Adger, N., J. Paavola, S. Huo and M. Mace (2006), *Fairness in Adaptation to Climate Change*, Cambridge MA: MIT Press.

Barnett, J. and N. Adger (2003), 'Climate dangers and atoll countries', *Climatic Change* 61, 321–37.

Borman, F. H. and S. Kellert (1991), *Ecology, Economics, Ethics: The Broken Circle*, New Haven: Yale University Press.

Bryner, G. (1999), 'Protecting humanity's future: threat, response, and debate', *Politics and the Life sciences* 18, 201–3.

Bryner, G. (2000), 'The United States: sorry – not our problem', in Lafferty, W. and J. Meadowbrook (eds), *Implementing Sustainable Development: Strategies and Initiatives in High Consumption Societies*, London: Oxford University Press, pp. 273–98.

Carter, N. (2004), *The Politics of the Environment*, Cambridge: Cambridge University Press.

Charles, L. (1987), *Toxic Wastes and Race in the United States: A National Report on the Racial and Socio-Economic Characteristics of Communities with Hazardous Waste Sites*, Baltimore: United Church of Christ Commission for Racial Justice.

Ciocirlan, C. and B. Yandle (2003), 'The political economy of green taxation in OECD countries', *European Journal of Law and Economics* 15, 203–18.

Clapp, J. and P. Dauvergne (2005), *Paths to a Green World: The Political Economy of the Global Environment*, Cambridge, MA: MIT Press.

Damro, C. and P. Luaces Méndes (2003), 'Emissions trading at Kyoto: from EU resistance to Union innovation', *Environmental Politics* 12, 71–94.

Davies, J. C. and J. Mazurek (1998), *Pollution Control in the United States: Evaluating the System*, Washington DC: Resources for the Future.

Dryzek, J. (1997), *The Politics of the Earth: Environmental Discourses*, Oxford: Oxford University Press.

Ekins, P. and T. Barker (2001), 'Carbon taxes and carbon emissions trading', *Journal of Economic Surveys* 15, 325–76.

Fiorino, D. (2001), 'Environmental policy as learning: a new view of an old landscape', *Public Administration Review* 61, 322–34.

Fischer, F. (2003), *Reframing Public Policy: Discursive Politics and Deliberative Practices*, New York: Oxford University Press.

Goldman, B. and L. Fitton (1994), *Toxic Waste and Race Revisited*, Baltimore: National Association for the Advancement of Colored People and United Church of Christ Commission for Racial Justice.

Haas, P. (1990), *Saving the Mediterranean: The Politics of International Cooperation*, New York: Columbia University Press.

Hajer, M. (1995), *The Politics of Environmental Discourse: Ecological Modernization and the Policy Process*, Oxford: Clarendon Press.

Hall, P. (1993), 'Policy paradigms, social learning and the state: the case of economic policy in Britain', *Comparative Politics* 25, 275–96.

Hossay, P. (2006), *Unsustainable: A Primer for Global Environmental and Social Justice*, London: Zed Books.

Howlett, M. and E. Lindquist (2004), 'Policy analysis and governance: analytical and policy styles in canada', *Journal of Comparative Policy Analysis* 6, 225–49.

Jacoby, W. (2006), *The Enlargement of the EU and NATO: Ordering from the Menu in Central Europe*, New York: Cambridge University Press.

Jasanoff, S. (1998), *The Fifth Branch: Science Advisers as Policymakers*, Cambridge MA: Harvard University Press.

Jasanoff, S. (2005), *Designs on Nature: Science and Democracy in Europe and the United States*, Princeton: Princeton University Press.

Jenkins-Smith, H. (1990), *Democratic Politics and Policy Analysis*, Pacific Grove CA: Brooks/Cole.

Jordan, A. (2005), *Environmental Policy in the European Union*, London: Earthscan.

Jordan, A., R. Wurzel and A. Zito (2003), ' "New" environmental policy instruments: an evolution or a revolution in environmental policy?' in Jordan, A., R. Wurzel and A. Zito (eds), *'New' Instruments of Environmental Governance? National Experiences and Prospects*, London: Frank Cass, pp. 199–224.

Jordan, A., R. Wurzel and A. Zito (2005), 'The rise of "new" policy instruments in comparative perspective: has governance eclipsed government?' *Political Studies* 53, 477–96.

Lafferty, W. and E. Hovden (2003), 'Environmental policy integration: towards an analytic framework', *Environmental Politics* 12(3), 1–22.

Lenschow, A. (ed.) (2002), *Environmental Policy Integration: Greening Sectoral Policies in Europe*, London: Earthscan.

Lévêque, F. (1996), *Environmental Policy in Europe: Industry, Competition and the Policy Process*, Cheltenham: Edward Elgar.

Linder, S. and G. Peters (1989), 'Instruments of government: perceptions and context', *Journal of Public Policy* 9, 35–58.

Oliver, M. and H. Pemberton (2004), 'Learning and change in 20th-century british economic policy', *Governance* 17, 415–41.

Pacala, S. and R. Socolow (2004), 'Stabilization wedges: solving the climate problem for the next 50 years with current technologies', *Science* 305, 968–72.

Richardson, J. and N. Watts (1986), *National Policy Styles and the Environment: Britain and West Germany Compared*, Berlin: Science Centre.

Romm, J. (2007), *Hell and High Water*, New York: William Morrow.

Sabatier, P. (1998), 'The advocacy coalition framework: revisions and relevance for Europe', *Journal of European Public Policy* 5, 98–130.

Schlosberg, D. (1999), *Environmental Justice and the New Pluralism*, Oxford: Oxford University Press.

Scott, R. (1995), *Institutions and Organizations*, Thousand Oaks: Sage.

Smith, A. (2000), 'Policy networks and advocacy coalitions: explaining policy change and stability in UK industrial pollution policy?' *Environment and Planning C* 18, 95–114.

US Environmental Protection Agency (2006), *Inventory of U.S. Greenhouse Gas Emissions and Sinks, 1990–2004*, http://www.epa.gov/climatechange/emissions/usinventoryreport.html [7 March 2008].

Victor, D. (2004), *Climate Change: Debating America's Policy Options*, New York: Council on Foreign Relations.

Vig, N. and M. Faure (eds) (2004), *Green Giants: Environmental Policies of the United States and the European Union*, Cambridge MA: MIT Press.

Vogler, J. (2005), 'The European contribution to global environmental governance', *International Affairs* 81, 835–50.

Ward, H. (2006), 'International linkages and environmental sustainability: the effectiveness of the regime network', *Journal of Peace Research* 43, 149–66.

Wettestad, J. (2005), 'The making of the 2003 EU emissions trading directive: an ultra-quick process due to entrepreneurial proficiency', *Global Environmental Politics* 5, 1–23.

Weyland, K. (2005), 'Theories of policy diffusion: lessons from latin american pension reform', *World Politics* 57, 262–95.

Wurzel, R. (2002), *Environmental Policy-making in Britain, Germany and the European Union: The Europeanisation of Air and Water Pollution Control*, Manchester: Manchester University Press.

Yandle, B. (1999), 'Public choice and the intersection of law and economics', *European Journal of Law and Economics* 8, 5–27.

Young, O. (ed.) (1997), *Global Governance: Drawing Insights from the Environmental Experience*, Cambridge MA: MIT Press.

Zito, A. (2000), *Creating Environmental Policy in the European Union*, Basingstoke: Macmillan.

3
The International Dimension of Climate Policy

Deborah Davenport

Introduction

As a tragedy of the global atmospheric commons (Soroos 2001), solutions to the problem of climate change will ultimately require global cooperation. Efforts at the national and EU levels to mitigate climate change are necessarily influenced by what happens on the international stage, so an understanding of climate politics at these levels also requires an understanding of this global context. This chapter, which is in part based on the analysis in Davenport (2006), provides a brief narrative history of developments at the international level through the negotiation of the Kyoto Protocol to the present day, along with a synopsis of the some of the main arguments in the academic literature on the international politics of climate change.

Negotiation of the climate regime: early developments

International climate politics has its roots in the World Climate Conference of 1979, at which scientists from 50 countries first reached consensus on the long-term significance of carbon dioxide levels in the atmosphere and established the World Climate Programme (WCP) under the auspices of the World Meteorological Organization (WMO), United Nations Environmental Programme (UNEP) and the International Council of Scientific Unions (ICSU). The WCP's broad objective was to determine how far climate could be predicted and the extent of human influence on climate (ICSU 2006) through long-term scientific research.

Under the WCP a series of international scientific meetings took place in Villach, Austria, during the early 1980s, culminating in 1985 in a consensus that a dialogue was needed between climate scientists and policy-makers. This led to two linked workshops, in Villach in September 1987 and Bellagio, Italy, two months later, at which the possible need for a convention to combat climate change was first voiced. This call came during an era of heightened environmental awareness in many industrialized countries prompted, *inter alia*, by the Bhopal disaster in 1984, the Chernobyl nuclear plant accident of 1986, several extreme weather events in the 1980s and the finding that the 1980s was at that time the hottest decade on record. Climate was thus one of several issues that contributed to forcing environmental issues onto the international political agenda (Schröder 2001).

In 1988 the WMO climate conference in Toronto interestingly couched climate change as a security issue in its title *International Conference of the Changing Atmosphere: Implications for Global Security* (Schabecoff 1988). The conference brought scientists and environmentalists together for the first time with policy-makers, although governmental officials attended in their own right rather than as delegates representing national interests. The participants agreed on a statement that included what became known as the 'Toronto target', calling on developed countries to reduce their CO_2 emissions by 20 per cent from 1988 levels by 2005 (WMO 1988).

The statement also called for the negotiation of a 'law of the atmosphere' convention, modelled on the UN Convention on the Law of the Sea, to address threats to the atmosphere, including climate change, ozone depletion and acid rain. However, the Montreal Protocol that formed the foundation for the ozone regime had been adopted less than two weeks before the 1987 Bellagio workshop and, ultimately, the framework convention-protocol approach adopted in the Montreal Protocol became the model for negotiations on climate change.

A separate but related process began in 1987, again under the auspices of the WMO and UNEP, in response to a US proposal to create an '*ad hoc* intergovernmental mechanism to carry out international co-ordinated scientific assessments of the magnitude, timing, and potential impact of climate change' (UNEP 1987: 76). With the creation of this body, which became the Intergovernmental Panel on Climate Change (IPCC), policy-makers first took control of addressing the climate issue. The IPCC was given a mandate to assess the latest scientific, technical and sociological literature and 'provide governments with a sound consensus of scientific evidence from which policy options can be developed' (United States Department of State 1988). These tasks were divided among three working groups covering (i) the risk of human-induced climate change; (ii) impacts, adaptation and vulnerability; and (iii) options for mitigation (IPCC 2007).

The IPCC has since issued four comprehensive assessments, the latest in 2007. The results of its first assessment report contributed to the Second World Climate Conference held in 1990. With a brief to recommend policy actions, the conference again brought together scientists, environmentalists and policy-makers, including ministers from 70 countries as well as six heads of state or government, leading to an official Ministerial Declaration. Although this statement was weaker than that issued by the scientists, particularly in that it did not refer to prospective reduction targets, it did identify principles for intergovernmental action that remain in force, including common but differentiated responsibilities for developed and developing countries; the concept of sustainable development; and the precautionary principle. It also contained the first official multilateral call for 'a framework treaty...and the necessary protocols – containing real commitments and innovative solutions' to address climate change in time for the planned UN Conference on Environment and Development (UNCED) in June 1992 (UNFCCC 1993).

The UN framework convention on climate change

Negotiations aimed at producing binding multilateral regulation to combat climate change were formally launched one month after the Second World Climate Conference, with the creation of the Intergovernmental Negotiating Committee on Climate Change (INC) by the UN General Assembly in December 1990. More than 150 states participated in the INC, which met five times before UNCED. Although the INC was mandated to draft a framework convention on climate change through broad consensus, it had to tackle numerous contentious issues identified at the 1990 climate conference (Carpenter *et al.* 1995a). Paterson and Grubb (1992) identified several major fault lines among the negotiating states as the early negotiations progressed:

- A North–South divide, particularly over how to share the burden of greenhouse gas reductions and how to use financial and technological transfers to assist developing countries to take action;
- A split between producers and exporters of oil and coal, and those who rely on fossil-fuel imports, over the question of reducing use of fossil fuels in order to reduce greenhouse gas emissions;

- A split between states that are relatively resilient to the impacts of climate change and those that are more vulnerable to extreme weather or rising sea levels and which lack the economic resources needed to adapt to such impacts.

The UN Framework Convention on Climate Change (UNFCCC) was signed at UNCED in Rio de Janeiro in 1992, but the most difficult questions were not resolved. Ultimately the UNFCCC contained only very weak language that all states could agree to in the context of wide divergences in interests. It did not establish firm targets for all Parties with timetables for achieving them, nor any implementing mechanisms or enforcement measures. Nevertheless, the UNFCCC achieved cooperation on several levels:

- It recognized that climate change is a real problem, despite the fact that its consequences are uncertain and long term in nature;
- Industrialized countries (OECD countries and Eastern European countries with 'economies in transition' to market-based systems) were denoted as 'Annex I countries' in the Convention and committed themselves 'specifically' to actions 'with the aim of returning individually or jointly to their 1990 levels [their] anthropogenic emissions of carbon dioxide and other greenhouse gases not controlled by the Montreal Protocol' (UNFCCC Article 4.2(b)). There was no timetable in the Convention for achieving this goal, however; this has been linked to strenuous US objections;
- To further this aim and to enable developing countries to make meaningful commitments for emissions reductions at a later time, the Convention required precise and regularly updated inventories of greenhouse gas emissions from all Parties;
- Parties committed themselves to take climate change into account in other areas of policy-making and to develop national climate programmes;
- Parties also accepted a binding principle of common but differentiated responsibilities on climate change, which put the heaviest burden on developed countries both as the source of most past and current greenhouse gas emissions and as the countries with the resources to help efforts elsewhere;
- To this end, Annex I countries committed themselves to providing financial support and sharing technology with less-developed countries.

The fact that the Convention was labelled a 'framework' document acknowledged the expectation that it would be amended or augmented over time to make intergovernmental efforts to address climate change more effective. This aim was pursued through the negotiation of the Kyoto Protocol.

The Kyoto Protocol negotiations

The UNFCCC entered into force less than two years after UNCED, and the first Conference of the Parties (COP-1) took place in 1995. A mandated COP-1 review of the adequacy of commitments produced three conclusions. First, most Annex I countries would not meet the UNFCCC's aim of lowering emissions by the implicit deadline of 2000 – this could be predicted even in the absence of a reduction target because emissions were expected to grow. There were also other weaknesses: the Convention contained no provision for actions after the turn of the century, and the goal of stabilizing emissions at 1990 levels was far too low to stem the problem of climate change, particularly as developing countries had made no commitment to go even this far. The Parties therefore reached a decision (known as the Berlin Mandate) which set in motion the negotiation of the Kyoto Protocol (Breidenich *et al.* 1998).

In making this decision, delegates had before them a draft protocol that had been produced months earlier by Trinidad and Tobago on behalf of the Association of Small Island States (AOSIS). AOSIS was formed in 1991 and comprises 43 small island and low-lying coastal countries that are highly vulnerable to rising sea levels. Reflecting this, the draft called for Annex I Parties to reduce their CO_2 emissions by at least 20 per cent from the 1990 baseline year by 2005, and to establish timetables for controlling emissions of other gases. The USA and Australia objected to this and the draft protocol's focus on CO_2 rather than all greenhouse gases. While the draft protocol did not call for developing countries to commit to emissions reductions, the huge divergence of interests between AOSIS and other developing countries, such as OPEC members and China, meant that it did not win much support in that group either (Carpenter *et al.* 1995b). Thus, the final Berlin Mandate contained, unsurprisingly, much weaker language than the AOSIS proposal.

The issue of developing countries' responsibilities did arise at COP-1, as a result of a German proposal to place commitments on developing countries according to their degree of industrialization. This idea was accepted by other OECD countries but consistently rejected by developing countries because nothing was offered in exchange. A major step forward appeared to be made at COP-2, probably influenced by the IPCC's Second Assessment Report in December 1995, which concluded that there was already a 'discernible human influence' on the climate and predicted an increase in average global temperatures of up to 6°C by 2100 if atmospheric CO_2 concentrations were not reduced (IPCC 1995). Another factor was the change in US administration since the adoption of the UNFCCC. The new Vice-President, Al Gore, was already associated with concern about climate change and the USA now stated willingness to negotiate firm targets and deadlines for reducing emissions (Brown 1995). The USA nevertheless insisted on 'flexibility' in how targets could be met, a concept it had introduced at the second meeting of the Ad Hoc Group on the Berlin Mandate (AGBM) in November 1995. These flexibility measures were an outgrowth of earlier calls by the USA for inclusion of all greenhouse gases, attention to 'sinks', such as forests, that absorb CO_2, and 'joint implementation' – the idea of allowing one country to finance emissions reduction in another country and claim these reductions against its own emissions target. Such flexible measures also included international emissions trading (Kyoto Protocol Article 17), whereby countries or companies are allocated credits, or 'rights to emit' within an agreed international cap, and then buy or sell credits depending on the relative costs of *not* reducing emissions and buying extra credits if necessary to cover these emissions, or reducing emissions and selling any unused credits. These flexibility measures reflected a new interest on the part of the USA in addressing the issue while forestalling the economic repercussions that might be expected given the US economy's high dependence on fossil fuels.

At COP-2, then, delegates were able to agree for the first time on a goal of negotiating legally binding targets, with the end of 1997 as a deadline. Although the OPEC countries, Russia and Australia objected, other delegates were able to finesse this by including the goal in a Geneva Ministerial Declaration, rather than in an official COP-2 decision. Even though not 'official', the majority support for this goal (now including the USA) was enough to turn the focus of negotiations towards its achievement by December 1997 (Schröder 2001).

In all the AGBM met eight times to negotiate the post-Convention stage of the climate regime prior to COP-3 in Kyoto. Delegates to AGBM meetings had to address three overarching thematic issues. First and foremost, Annex I countries had to negotiate emissions targets to which they would be willing to commit themselves. Second, agreement on targets would depend on simultaneous acceptance of the US-proposed 'flexibility mechanisms', which raised many new and complex questions. Finally, there was the question of new commitments by non-Annex I countries, which brought with it the question of incentives, including financial or technological assistance.

On the issue of commitments, negotiators of the Kyoto Protocol generally followed the Montreal Protocol model in an attempt to agree specific emissions reductions for six greenhouse gases – CO_2, methane, nitrous oxide, hydrofluorocarbons (HFCs), perfluorocarbons (PFCs), and sulphur hexafluoride (SF_6) – with a timetable for achievement by 2008–2012. This ultimately led to agreement by developed country parties to reduce their overall emissions of these gases by at least five per cent below 1990 levels for the commitment period 2008–2012 (Kyoto Protocol Article 3.1).

This commitment is weak in numerous ways. First, it is inadequate in comparison with scientific assessments of what is required, especially given the Toronto target of a 20 per cent reduction of CO_2 emissions by 2005, with a 50 per cent cut in the long term (Lomborg 2001; Schröder 2001), much less the 60 per cent long-term reductions requirement called for in the IPCC's First Assessment Report in 1990. Second, Annex I countries failed to find a general accepted formula for determining their 'quantified emissions limitation and reduction objectives' (QELROs), so the QELROs ultimately listed in an annex of the Protocol were the result of political expediency: each committing country chose for itself the emissions reductions it would attempt to achieve over the period 2008–2012. While many countries chose targets in the seven to eight per cent range, Australia, Iceland and Norway only committed to reduce their emissions *growth*, not to reduce emissions (Annex B). Moreover, the fact that Australia did not ratify the Protocol until December 2007 arguably meant a ten-year delay in acting on commitments made in 1997 (see Bailey and Maresh, this volume).

On the second issue, the Kyoto Protocol ultimately included all the 'flexibility mechanisms' the USA had demanded. Calculation of changes in emissions is to take account of changes in greenhouse gas removals by sinks resulting from direct human-induced land-use change and forestry activities since 1990 (Article 3.3). 'Supplemental' emissions trading between parties is allowed under Article 17, and transfer of credits resulting from projects aimed at reducing emissions or enhancing removals of greenhouse gases by sinks is also allowed between Annex I parties under Article 6 (so-called Joint Implementation). Further flexibility was obtained through agreement to allow 1995, rather than 1990, to be the base year for HFCs, PFCs and SF_6 (Article 3.8).

Finally, QELROs were only made by Annex I countries. The UNFCCC and Kyoto Protocol established mechanisms intended to assist implementation and encourage developing country participation, although both catered heavily to donor interests. The UNFCCC refers to a 'financial mechanism' which was entrusted to the already-established and World Bank-controlled Global Environment Facility. The Protocol also establishes the Clean Development Mechanism, which, unlike the Montreal Protocol's Multilateral Fund, is not a fund but rather is supposed to assist countries 'as necessary' (Article 12.6) in funding project activities from other parties, which would then be allowed to use certified emission reductions accruing from projects towards compliance with their own QELROs commitments under Article 3 (Article 12.3(b)). The implications of this lack of commitment by developing countries were significant, given that the world's second and fifth largest CO_2 emitters, China and India, were not Annex I parties and were therefore not covered by the Protocol's commitments to reductions.

From Kyoto 1997 to Bali 2007

Despite the fact that the negotiation of the Kyoto Protocol lasted 24 hours longer than the time allotted for COP-3, of necessity it left many specifics to be negotiated later. In addition, some of the troublesome ideas that had been addressed in Kyoto remained subjects of contention, such as the question of how far the flexibility mechanisms could substitute for domestic action to meet emissions targets, and the issue of equity with regard to developing countries. It took four years to reach agreement on this 'operationalization' of the Protocol.

Meanwhile, the US Senate had unanimously passed a non-binding resolution shortly before COP-3 that the USA should not sign any protocol that mandated new emissions commitments for Annex I parties unless it 'also mandate[d] new specific scheduled commitments for developing country parties within the same compliance period' (US Senate Debate 2004: 117–29). Although the USA signed the Protocol, the resolution meant that President Clinton never felt able to send it to the Senate for ratification. In 2001 the new President Bush went further and 'withdrew' the US signature, after which time the USA no longer participated in debates concerning the details of the Protocol in any meaningful way. This prompted efforts to weaken the agreement enough to entice the USA back into the fold, which in turn shifted the politics of ratification in favour of other Annex I countries that also preferred a weaker commitment, including Russia, Canada, Japan and Australia, as it meant that their ratification was now required to ensure the Protocol was ratified by Annex I parties accounting for at least 55 per cent of CO_2 emissions in 1990 and, thus, could enter into force (Porter *et al.* 2000).

After a stalemate at COP-6 in the Hague in late 2000 and the change in US administration in early 2001, this 'anti' coalition won a series of concessions at COP-6 (Part 2) in Bonn the following July, including low eligibility requirements for the use of flexibility mechanisms and broader interpretation of provisions on emissions reduction credits for changes in land use and forestry activities to include forest management, cropland management, grazing land management and revegetation (Hanks *et al.* 2001). The concessions also included a weak compliance system, with sanctions comprising a reduction in a country's allowed emissions during the envisaged second Protocol commitment period (post-2012). This, however, is meaningless if there is no second commitment period, and the post-2012 future is still far from certain.

These compromises resulted in commitments that, according to news reports at the time, would reduce emissions levels by only two per cent below their 1990 levels by 2012, rather than the 5.2 per cent agreed in Kyoto; Greenpeace pejoratively labelled the Bonn agreements 'Kyoto lite' (Kirby 2005). At COP-7 in Marrakesh, later that year, these agreements were formalized into a set of decisions detailing rules and procedures for operationalizing the Kyoto Protocol, which became known as the Marrakesh Accords. The Accords were not fully finalized and adopted until the first Meeting of the Parties to the Kyoto Protocol (MOP) in November 2005, which took place in conjunction with COP-11. That this meeting took place at all was remarkable, because until November 2004 many doubted whether Kyoto would ever enter into force. Without the USA, the Protocol had to be ratified by all other Annex I countries, many of whom were still reluctant. The EU, influenced by Germany as the strongest proponent of the climate change regime, had enough leverage to offer Russia an adequate sweetener: EU support for Russian entry into the World Trade Organization. A deal was thus made whereby Russian ratification brought the required amount of developed country emissions to be covered by the Protocol above the 55 per cent threshold. The Kyoto Protocol came into effect on 16 February 2005, three months after Russia's ratification.

Immediately upon Kyoto's entry into force, the next stage of global negotiations on what would happen after 2012 had to be tackled. The Kyoto Protocol itself mandated that consideration of such further commitments be initiated at least seven years before the end of the first commitment period, in other words by 2005 (Article 3.9). At COP-11/MOP-1, with the new momentum given to the Kyoto Protocol, the USA pursued the only path open to it. Given that it was not a party to the Protocol, it would not negotiate future commitments under it but, instead, insisted on a separate track for negotiations of an entirely new instrument or protocol under the UNFCCC. While this 'two track' approach engendered much debate, it was eventually accepted as the only way to move forward and bring countries such as the USA on board. As the USA was the world's largest economy and largest single emitter of CO_2 (closely followed by China), this was a key aim for many. In the end, parties to the Kyoto Protocol (COP-11/MOP-1) created an Ad Hoc Working Group on Further Commitments for Annex I parties under the Kyoto Protocol (AWG) and established a separate Dialogue on long-term cooperation under the UNFCCC. A third process was also begun, to review commitments under Kyoto mandated under Article 9.

By COP-13/MOP-3 in Bali in December 2007 a decision was due on how to proceed in relation to negotiating a new instrument for the post-2012 era. This put renewed pressure on negotiators and ultimately led to a late-hour agreement on a 'Bali Roadmap', which outlines a two-year process to agree a post-2012 climate regime by 2009. The group eventually agreed on keeping Convention and Protocol work under separate tracks, while formalizing the Convention dialogue track into a second *ad hoc* working group. Interestingly, the decision establishing this new *ad hoc* working group sets out for the first time a negotiating agenda that contemplates mitigation actions for both developing and developed countries and avoids referring to Annex I and non-Annex I countries, thereby giving more flexibility for considering countries' contributions to a future agreement.

Frameworks for analysis

Numerous writers have attempted to use theoretical analysis both to explain the unfolding process of climate politics and to guide policy-makers towards more effective actions. In the early 1990s much attention was directed towards comparisons with the apparent success story of the Montreal Protocol (for instance, Benedick 1991). There were numerous similarities between the two issues, including the fact that they both deal with management of the atmospheric commons. Ultimately, however, analysts realized that climate change presented many more policy challenges than did ozone depletion because whole economies were dependent on fossil fuels, unlike the more isolable ozone-depleting substances. This was especially the case for the USA, with its immense and relatively inefficient fossil-fuel use. Whereas the USA championed the Montreal Protocol, it was a laggard on climate change (Andresen and Agrawala 2002). This had profound implications for reaching an effective international agreement.

More recent analyses have focused on why global climate politics have been so problematic and what can be done about it (Soroos 2001; Sprinz and Weiβ 2001). One fruitful path has been taken by several works seeking to identify why configurations of interests are formed, the extent to which common interests are held, reasons behind interest divergence or convergence, how interests shift, and how these all affect the outcomes of negotiations. This literature may be dated to Haas' (1992) theory on epistemic communities: scientific networks that produce and use common scientific knowledge to influence policy-makers, and to Sprinz and Vaahtoranta's (1994) works on interest-based explanations for international environmental policy. Haas' claims about the possibility of leadership and influence from epistemic communities when scientific consensus exists seemed to explain much about the creation of the ozone regime but left open the question of why this was not replicated in the climate case. At the same time, Haas' claims about the influence of epistemic communities have been sharply criticized, particularly for their lack of clarity about how the knowledge community and those in power influence each other. Moreover, numerous studies have questioned the ability of scientific communities to influence all negotiations in the same way, given the existence of adversarial science (Susskind 1994) and the fact that increasing scientific certainty is not necessarily associated with greater interest in cooperative action unless new information leads to predictions at least as threatening as those made earlier (Bauer 2006).

Sprinz and Vaahtoranta (1994) attempt to explain cross-national variance in support for international environmental regulation using two interest-based factors, *ecological vulnerability* (the environmental threat a state faces) and *economic capacity* (the economic abatement costs for individual countries). However, they do not investigate whether an agreement is likely when sets of parties to a negotiation hold differing interests. This likelihood might depend on other factors, such as the strength of the states pushing for regulation. Andresen and Agrawala (2002) further this categorization in their study of leaders, pushers and laggards in the climate regime during the 1990s, but, again, without reaching conclusions on how cooperation might be enhanced.

Others, such as DeSombre (2000) and Barrett (2003), more specifically address the implications of differing interests for the development of international environmental regimes. DeSombre examines international environmental cooperation as the result of the internationalization of domestic policy. DeSombre first targets which domestic policies are chosen for internationalization, and then the extent to which states wanting to internationalize national policy succeed in convincing other states to adopt similar policies. For a state to be motivated to support or push for international environmental policy there must be both strong environmental as well as economic incentives for domestic actors, but for DeSombre the economic interests involved are broader than simple abatement costs. If industry is already affected by the costs of meeting domestic regulations it has an incentive to push for their internationalization to make other states bear similar environmental protection costs; alternatively it will push for international regulations to exclude goods produced by states not subject to those costs. DeSombre's analysis does not focus on climate change but is easily extrapolated to that issue.

Barrett's (2003) work addresses these questions more directly in relation to climate politics. Starting with a prisoner's dilemma model, Barrett highlights the need to restructure incentives to enforce participation and compliance, in order to achieve effective climate treaty design. This framework, however, assumes common interests and forces a strained interpretation of all negotiations as lying within the scope of enforcement issues: 'Recognizing that it is in their joint interests to [abate pollution], we might suppose that the two countries will negotiate an agreement which alters the payoffs in such a way that each state's own interests compel it to play "abate" (in other words, to cooperate)' (Barrett 2003: 62). Barrett conceptualizes the main shortcoming of the Kyoto Protocol as enforcement, because negotiators did not deal with this issue in a timely way because of a mistaken view among negotiators that it could be addressed later or, as Kyoto participants told him, that the issue was put off because 'you cannot solve every aspect of this problem in one stroke' (2003: 360–2). This, however, does not address the question of whether the enforceability of the commitments made might run counter to the interests of powerful actors.

For an agreement to be self-enforcing requires that all parties benefit more from an agreement than from the status quo, a point already made by Keohane (1984), who also appears to assume a common interest in an agreement that can be effectively enforced. Neither Barrett nor Keohane come fully to grips, however, with the fact that the effort to design a treaty incorporating such incentives is dependent on the bargaining process, or the fact that bargaining depends on leverage, which has much to do with the power structure involved.

Davenport (2006) attempts to show that in an asymmetric international system, effective agreement to ameliorate climate change or any other global environmental problem depends upon whether the lead state in the system (the USA) is willing to bear the costs of manipulating other states' preferences towards effective agreement. This depends on the costs and benefits of agreement to the lead state, and Davenport provides a typology of these potential benefits and costs to clarify the leader's true interests on a particular issue such as climate change. Potential benefits include the obvious environmental benefits but also the avoidance of costs or the positive economic benefits that industry may gain from international regulation on particular environmental issues. Costs entail the potential cost of halting regulated activities or activities dependent on a banned substance, the cost of developing substitutes where possible and the cost of manipulating other countries' preferences. Thus, the effectiveness of the future global climate regime depends in large measure on shifting perceptions of these costs and benefits within the USA. To this list might be added, in the near future, the costs that may be linked to climate change as a security threat, about which concern has recently begun to swell within the USA and elsewhere (Brown *et al.* 2007; CNA 2007; United Nations Security Council Department of Public Information 2007).

In a similar vein, Bang *et al.* (2007) attempt to judge the likely configuration of the post-2012 climate regime based on perceived interests within the USA. Given that no major new commitments may be expected from any country unless the USA also comes on board, it appears likely that any post-2012 regime will entail a new instrument under the UNFCCC rather than an amended Kyoto Protocol, as per the 'Dialogue' path pushed by the USA, and that this could well entail an 'exten[sion of] US climate policy to other countries' (Bang *et al.* 2007: 1289). For both Davenport and Bang *et al.*, therefore, the key question for international climate governance is how interests within the USA ultimately shape national US climate policy and how they may be influenced themselves. The present work addresses this question through detailed comparative analysis of climate politics in the USA and other developed countries.

References

Andresen, S. and S. Agrawala (2002), 'Leaders, pushers and laggards in the making of the climate regime', *Global Environmental Change* 12, 41–51.

Bang, G., C. Froyn, J. Hovi and F. Menz (2007), 'The United States and international climate cooperation: international "pull" versus domestic "push"', *Energy Policy* 35, 1282–91.

Barrett, S. (2003), *Environment and Statecraft: The Strategy of Environmental Treaty-making*, New York: Oxford University Press.

Bauer, S. (2006), 'Does bureaucracy really matter? The authority of intergovernmental treaty secretariats in global environmental politics', *Global Environmental Politics* 6, 23–49.

Benedick, R. (1991), *Ozone Diplomacy: New Directions in Safeguarding the Planet*, Cambridge: Harvard University Press.

Breidenich, C., D. Magraw, A. Rowley, and J. Rubin (1998), 'The Kyoto Protocol to the United Nations Framework Convention on Climate Change', *American Journal of International Law* 92, 315–31.

Brown, P. (1995), 'Way open for cuts in greenhouse gases: breakthrough clears a path for hard bargaining on targets and implementation timetables', *The Guardian* 8 April, 12.

Brown, O., A. Hammill and R. Mcleman (2007), 'Climate change as the "new" security threat; implications for Africa', *International Affairs* 83, 1141–54.

Carpenter, C., P. Chasek and S. Wise (1995a), 'Summary report on the first conference of the parties to the framework convention on climate change: 28 March–7 April 1995', *Earth Negotiations Bulletin* 12.

Carpenter, C., P. Chasek, L. Goree and S. Wise (1995b), 'Summary of the Eleventh Session of the INC for a Framework Convention on Climate Change: 6–17 February 1995,' *Earth Negotiations Bulletin* 12.

CNA (The CNA Corporation) (2007), *National Security and the Threat of Climate Change*, securityandclimate.cna.org/ [8 February 2007].

Davenport, D. (2006), *Global Environmental Negotiations and US Interests*, New York: Palgrave Macmillan.

DeSombre, E. (2000), *Domestic Sources of International Environmental Policy: Industry, Environmentalists, and US Power*, Cambridge: MIT Press.

Haas, P. (1992), 'Introduction: epistemic communities and international policy coordination', *International Organization* 46, 1–37.

Hanks, J., L. Schipper, M. Sell, C. Spence and J. Voinov (2001), 'Summary of the resumed sixth session of the Conference of the Parties to the UN Framework Convention on Climate Change: 16–27 July 2001', *Earth Negotiations Bulletin* 12.

ICSU (International Council of Scientific Unions) (2006), *ICSU and Climate Science: 1962–2006 and Beyond, From GARP to IPCC*, www.icsu.org/10_icsu75/PDF/Climate_Change.pdf [26 August 2007].

IPCC (Intergovernmental Panel on Climate Change) (1995), *Second Assessment Synthesis of Scientific-Technical Information Relevant to Interpreting Article 2 of the UN Framework Convention on Climate Change*, Geneva: IPCC.

IPCC (Intergovernmental Panel on Climate Change) (2007), www.ipcc.ch/ [25 August 2007].

Keohane, R. (1984), *After Hegemony: Cooperation and Discord in the World Political Economy*, Princeton: Princeton University Press.
Kirby, A. (2005), 'The Bonn deal: winners and losers', BBC News, 23 July, news.bbc.co.uk/1/hi/sci/tech/1452903.stm [12 October 2007].
Lomborg, B. (2001), *The Skeptical Environmentalist: Measuring the Real State of the World*, Cambridge: Cambridge University Press.
Paterson, M. and M. Grubb (1992), 'The international politics of climate change', *International Affairs* 68, 293–310.
Porter, G., J. Brown and P. Chasek (2000), *Global Environmental Politics*, Boulder: Westview Press.
Schabecoff, P. (1988), 'Norway and Canada call for pact to protect atmosphere', *New York Times*, 28 June.
Schröder, H. (2001), *Negotiating the Kyoto Protocol: An Analysis of Negotiation Dynamics in International Relations*, Münster: Lit Verlag.
Soroos, M. (2001), 'Global climate change and the futility of the Kyoto process', *Global Environmental Politics* 1, 1–9.
Sprinz, D. and T. Vaahtoranta (1994), 'The interest-based explanation of international environmental policy', *International Organization* 48, 77–106.
Sprinz, D. and M. Weiβ (2001), 'Domestic politics and global climate policy', in Luterbacher, U. and D. Sprinz (eds), *International Relations and Global Climate Change*, Cambridge: MIT Press, pp. 67–94.
Susskind, L. (1994), *Environmental Diplomacy: Negotiating More Effective Global Agreements*, Oxford: Oxford University Press.
UNEP (United Nations Environmental Programme) (1987), 'Report of the Governing Council on the work of its fourteenth session, 8–19 June 1987', *General Assembly Official Records* Second Session Supplement No. 25 (A/42/25).
UNFCCC (United Nations Framework Convention on Climate Change) (1993), *The Second World Climate Conference*, unfccc.int/resource/ccsites/senegal/fact/fs221.htm [9 September 2007].
United Nations Security Council Department of Public Information (2007), 'United Nations Security Council hold first ever debate on impact of climate change on peace, security, hearing over 50 speakers', *UN Security Council 5663rd meeting,* 17 April 2007, www.un.org/News/Press/docs/2007/sc9000.doc.htm [8 February 2007].
United States Department of State (1988), 'International cooperation to study climate change', Statement by Richard J. Smith, Acting Assistant Secretary for Oceans and International Environmental and Scientific Affairs, before the House Foreign Affairs Committee on March 10, 1988, *US Department of State Bulletin*, June.
US Senate Debate over the Byrd-Hagel Resolution, July 25, 1997 (2004), in Victor D. (ed.), *Climate Change: Debating America's Policy Options*, Appendix A, Council on Foreign Relations, pp. 117–29.
WMO (World Meteorological Organization) (1988), *World Conference on the Changing Atmosphere – Implications for Global Security, Toronto 1988*, Conference Proceedings, Report No. 710, Geneva: WMO.

Part II
Analysis

4
The European Union and the Politics of Multi-Level Climate Governance

Chad Damro and Donald MacKenzie

Introduction

While the European Union (EU) is a prominent player in the politics of climate change, it is neither a state nor an international organization in the traditional sense. Rather, it operates as a proactive and authoritative regional collective of affluent democracies that can influence policy-making in significant ways at the regional and international levels. This unique position also means that EU policy-making is subject to multiple pressures from both these levels. Despite – and possibly because of – this, the EU proudly promotes its collective efforts as an exemplar of how to tackle climate change through a combination of international and regional commitments.

This chapter begins by discussing the domestic and international foundations of EU climate policy. It then explores political analysis conducted in this area, including explanations for developments in climate policy at the EU level. Next, it identifies a number of international obstacles to EU climate policy, and domestic and regional obstacles to its Emissions Trading Scheme. Particular attention is given to emissions trading, rather than the EU's initiatives on renewable energies, biofuels and vehicle emissions, because emissions trading is widely regarded as the mainstay of the EU's climate strategy, now and into the future. It also exemplifies many of the generic political tensions that exist within EU climate policy. The chapter concludes by identifying political strategies available to the EU for overcoming these obstacles and by arguing that, despite the multiple domestic and international pressures facing the EU, it seems certain to play a sustained and active role in this policy area.

EU climate policy: domestic and international foundations

The EU's extensive authority in environmental policy is especially noteworthy given that environmental policy was not included in the primary legislation (treaties) of the EU until the 1986 Single European Act. As the 27 member states have pooled sovereignty in environmental policy, the Union has developed the legal and political capacity to play a significant role in international environmental policy-making and to determine domestic climate change legislation. For simplicity, this study refers to the 'EU' throughout, despite legal distinctions that exist between the EU and European Community (EC) in this policy area. The term 'EC' will be used only when necessary for legal clarity and when cited in secondary sources.

At the international level, the EU has been an active participant in United Nations Framework Convention on Climate Change (UNFCCC) negotiations since their beginning. The EU and its member states actively promoted the Kyoto Protocol and the 2002 Marrakech Agreement and were rewarded for their efforts in 2005 when enough countries ratified the Protocol for it to enter into force. A contentious international priority for the EU during these negotiations has been the establishment of binding emissions reduction targets within set timeframes for Annex I countries. Despite shifting positions and fluctuating influence during the nearly decade-long UNFCCC negotiations – notably at the Sixth Conference of Parties in The Hague (Grubb and Yamin 2001) – the Union is now often described as a 'leader' or 'frontrunner' in international climate policy-making (Gupta and Grubb 2000; Gupta and Ringius 2001; Andresen and Agrawala 2002; Christiansen and Wettestad 2003; Zito 2005; Skodvin and Andresen 2006).

As the EU has established itself in this area, its internal policy actors have had to navigate a unique landscape of regional institutions. Space constraints prevent a detailed review of the EU's various internal decision-making bodies, which include the European Commission, European Council, Council of Ministers, European Parliament and European Court of Justice (McCormick 2001; Jordan 2005; Lenschow 2005; Jordan and Schout 2006), or its decision-making procedures. However, it is worth noting that the Commission holds primary responsibility for proposing new policies, under the broad strategic guidance given by the European Council of Heads of State and more specific requests from relevant Councils of Ministers, and for ensuring that member states implement EU laws properly. Decisions in most areas of environmental policy on whether to accept or veto Commission proposals are made by the Council of Ministers in co-decision with the European Parliament. Measures affecting taxation powers, choices on the structure of energy supply and most areas of land-use planning all require unanimous Council approval, whereas qualified majority voting is generally applied to other relevant policy areas.

The Commission has undertaken a number of EU climate-related initiatives since 1991, when it issued the EU's first strategy to limit CO_2 emissions and improve energy efficiency. This strategy included measures to promote renewable energy, the securing of voluntary commitments by automobile manufacturers to reduce CO_2 emissions (upgraded to mandatory targets in 2008) and proposals for common taxes on energy products. The Council of Environment Ministers then asked the Commission to develop priority actions and policy measures, which resulted in the launch of the European Climate Change Programme (ECCP) in June 2000. The ECCP has acted as the Commission's main instrument to identify and develop an EU strategy to implement the Kyoto Protocol. The negotiations over the first ECCP involved a number of stakeholder groups, including representatives from the Commission's Directorates-General, member states and industry and environmental groups. The political influence exercised by these different actors often varies across the different issues and instruments under discussion. Likewise, political influence and the likelihood of policy change often varies with the specific constellation of member states actively involved, in particular the positions taken by environmental leaders and laggards within the Union (Börzel 2000; Lenschow 2005). A case in point is the failed proposal for a common EU carbon/energy tax which, as a measure that would have conferred taxation powers on the EU, required unanimous Council support to come into force but was opposed by various member states on economic or national sovereignty grounds. The compromise solution was relatively lax common minimum duties on a range of energy products.

As is shown in Table 4.1, the ECCP has generated a considerable volume of EU-level legislation, mainly in the form of directives that the member states are legally bound to transpose into national laws. According to the Commission's accounting, the EU has introduced over 30 climate change initiatives since 2000.

The EU launched its second ECCP in October 2005. This is designed to run in close cooperation with a wide range of stakeholders and is organized around several working groups tasked with reviewing ECCP I (with five subgroups: transport, energy supply, energy demand, non-CO_2 gases and agriculture) and the EU's Emissions Trading Scheme, as well as

Table 4.1 The European Climate Change Programme

Measure	Reduction potential (millions of tonnes of CO_2 equivalent) EU-15, 2010	Entry into force	Starting to deliver
EU emissions trading scheme	–	2003	2005
Links to Joint Implementation and Clean Development Mechanism	–	2004	2005–8
Directive on promotion of electricity from renewable energy sources	100–125	2001	2003
Directive on promotion of combined heat and power	65	2004	2006
Directive on energy performance of buildings	35–45	2003	2006
Directive on promotion of transport biofuels	35–40	2003	2005
Landfill directive	40	1999	2000
Vehicle manufacturer voluntary commitment (since replaced by mandatory targets)	75–80	1998	1999
Energy labelling directives	20	1992	1993
Biomass action plan	–	2005	2006

Source: Delbeke (2006: 6).

exploring possible climate measures relating to aviation, automobiles, carbon capture and storage and adaptation to climate change.

The Commission also organizes its work around the EU Environmental Action Programmes (EAP), which set out the framework and strategic priorities for EU environmental policy. These are non-binding frameworks that establish agendas, but the individual regulatory interventions that follow are still subject to political negotiations on a case-by-case basis. The most recent Sixth EAP runs from 2002–2012 and includes four priority areas: climate change; nature and biodiversity; environment and health; and natural resources and waste. The earlier Fifth EAP (1993) also included climate change among its themes.

The EU's ambitious position on greenhouse gas emissions reductions was clearly elaborated by the European Council meeting of Heads of

State and Government held in March 2007, where it was agreed that the EU would cut its emissions to at least 20 per cent below 1990 levels by 2020. In addition, the EU committed to cutting 'its emissions to 30 per cent below 1990 levels by 2020 provided that, as part of a global and comprehensive post-2012 agreement, other developed countries commit to comparable reductions and advanced developing countries also contribute adequately to the global effort according to their respective capabilities' (European Commission 2007: 9). The EU intends to achieve these reductions through the measures agreed in the ECCPs and 'new measures included in an integrated climate and energy strategy' (European Commission 2007: 9). The Commission released the first wave of proposals in January 2008, which included a major expansion in the stringency and scope of the EU Emissions Trading Scheme (EU ETS).

Additional climate change measures include further funding for research and technological development. The EU's Seventh Framework Programme for Research and Development (2007–2013) has an increased budget of €8.4 billion allocated for environment, energy and transport. This programme is designed to assist the 'soonest possible deployment of clean technologies as well as further strengthening knowledge of climate change and its impacts' (European Commission 2007: 12). The EU is committed to increasing this research budget further after 2013.

The EU's flagship policy to combat climate change is, however, undoubtedly the EU ETS (Watanabe and Robinson 2005). The establishment of the EU ETS demonstrates how the Union can operate as an authoritative regional point of interaction between the national and the international levels. At the national level, the EU ETS now covers roughly half of the EU's CO_2 emissions. At the international level, it represents a case in which the EU changed its position and now seems to be demonstrating international leadership by example. In operational terms, the promise of the EU ETS seems positive, but questions remain about the modalities of emissions trading, the competing interests engaged in emissions trading, and the actual abatement that will result from emissions trading processes. The EU ETS is also likely to serve as a future linking system to other national, regional and international emissions trading schemes (Oberthür 2006; Legge 2007). For example, the EU ETS recognizes Clean Development Mechanism and Joint Implementation credits for emissions reductions financed by one country in another country, up to certain agreed limits set at national level, as equivalent emissions allowances that can be used within the scheme. A brief summary of the key points of the EU ETS is provided in Box 4.1.

Box 4.1 Key features of the EU emissions trading scheme

The pilot phase of the EU ETS ran from 2005 to 2007. Phase 2 (2008–2012) corresponds to the assessment period for the Kyoto Protocol. Phase 3 (2013–2020) is set to improve and extend emissions trading, taking into account future international agreements following the expiry of the Kyoto Protocol in 2012.

The EU ETS is a 'cap and trade' scheme. First, the regulator sets an overall cap on emissions and allocates allowances to greenhouse gas producers. Allowances are then traded on an open market to provide financial incentives for emissions reduction. Under the first two phases of the EU ETS, large emitters from the energy, ferrous metals, minerals and pulp and paper sectors were given annual allocations of allowances, with each allowance corresponding to one tonne of CO_2. These were based on National Allocation Plans drawn up by member state governments and agreed with the European Commission, rather than through a single EU cap.

During Phase 1, 95 per cent of allowances could be allocated free of charge, the remainder being auctioned. The maximum proportion of free allocations reduced to 90 per cent in Phase 2. Allowances can be traded freely between firms, sectors and member states, but target installations must submit annual returns to their national verification body each year cancelling permits corresponding to their actual emissions. The penalty for submitting insufficient allowances during Phase 1 was €40 per excess tonne of CO_2; this rose to €100 in Phase 2.

In January 2008, the Commission released its plans for Phase 3 of the EU ETS. These include proposals for an EU-wide emissions cap (as opposed to member-state determined caps), 80 per cent auctioning of allowances, the inclusion of other greenhouse gases and sectors (notably aviation) within the scheme, new rules on the use of credits from emissions-reduction projects in third countries and flexibility provisions to take into account the differing abatement capabilities and development needs of richer and poorer member states (European Commission 2008).

Despite its unique and complex political arrangements, the EU has engaged actively in the initiation, institutionalization and implementation of a variety of climate-related policies. Because of its unique nature, the EU has had to develop a system of governance capable of channelling various domestic and international pressures to its advantage. The result has been a comprehensive ECCP that includes emissions trading and international recognition as an environmental leader.

Political analysis of EU climate policy

EU climate policy has generated a vast amount of practical and academic debate and research in recent years. The practical debate and analysis has engaged citizens, media, public authorities, the private sector and non-governmental organizations (NGOs) (Mazey and Richardson 1992; Michaelowa 1998) as well as policy institutions such as the Institute for European Environmental Policy, Ecologic, Centre for European Policy Studies and European Environmental Bureau. The EU holds a variety of stakeholder consultation workshops on issues such as the Green Paper on Adapting to Climate Change in Europe, and its ECCPs have also benefited from the input of such stakeholder groups. In 1990 the EU made an internal institutional stride into this public debate when the Council approved the creation of the European Environment Agency (EEA). The key role of the EEA is information provider and analyst, rather than participant in policy-making, and it boasts a membership of over thirty countries, including non-EU states like Turkey and Switzerland.

In addition to civil society actors and the EEA, numerous academics have weighed in with analyses of competing policy options as well as of the technical and economic implications of EU climate policy. For example, scholars have analyzed the modalities and politics of burden-sharing (Oberthür 2006), national allocation plans (Betz *et al.* 2006), issues surrounding the auctioning of emissions permits (Mandell 2005; Hepburn *et al.* 2006), challenges to the EU ETS (Grubb and Neuhoff 2006), and options for the EU's long term-strategies and goals in climate policy (Winne *et al.* 2005).

The academic literature on the politics of climate change also covers a number of international and regional issues related to EU climate policy and the linkages across different levels of analysis. The decision to establish the EU ETS provides a useful example of such cross-level linkages. The creation of the world's largest and most comprehensive emissions trading scheme in 2003 was a major innovation, with significant costs in terms of time and other resources. Add to this the international uncertainty surrounding the Kyoto Protocol when the EU began formulating the EU ETS, and the Union's decision to move forward with the initiative seems particularly puzzling and risky. Many factors from the national, EU and international levels have had an impact on this decision, which several studies have tried to disentangle.

First, studies have explored the EU's motivations for adopting the idea of emissions trading after initially resisting it in international negotiations. Damro and Luaces-Mēndez (2003) argue that the EU did so as part of a process of policy learning from USA experiences with similar domestic schemes. Woerdman (2004) moves beyond policy learning to argue from a path-dependence approach that the shift occurred as the result of internal and external pressures to maintain climate leadership. Cass (2005) argues that the EU's advocacy of emissions trading is best understood as the result of shifting 'frames' of debate that allowed the Union to overcome domestic obstacles that had previously prevented support for other market-based mechanisms.

Other studies have focused on the specific reasons why the EU issued its 2003 directive establishing the EU ETS. Wettestad (2005) tends to emphasize the central role played by the Commission in utilizing climate science and emissions trends to overcome veto points, while Oberthür (2006) and Oberthür and Tänzler (2007) emphasize the causal role of international regimes. The sum total of these scholarly efforts suggests that explanations of the EU ETS need to consider a significant causal role for domestic *and* international factors.

It is worth identifying briefly some important institutional and other pressures from different levels that help to explain the EU ETS. At its most basic level, the EU ETS arose from the UNFCCC and the resulting Kyoto commitments. Early in the negotiations, the EU resisted emissions trading in favour of more command-and-control regulatory and taxation schemes. By contrast, the USA was the primary driver of this instrument based on its experience with domestic sulphur dioxide trading (Christiansen and Wettestad 2003; Damro and Luaces-Mēndez 2003). As the EU gradually changed its position, the USA reduced its commitment to the Kyoto Process as President Clinton decided not to send the Protocol to a Senate that publicly opposed ratification, and President Bush repudiated the Protocol altogether in March 2001 (Lisowski 2002; Steurer 2003).

The differing EU and US positions were a point of contention from the outset of the negotiations. As Sbragia (1998: 299) points out, as early as 1992 EU Finance Ministers insisted that any EU carbon tax be implemented only on the condition that the USA and Japan acted in kind. Japan agreed on condition that the USA enact some kind of carbon tax. The Clinton administration refused. The EU's gradual acceptance of emissions trading allowed for compromise and created an opportunity for progress in the negotiations. Some of the change in the EU position can certainly be attributed to an international process of policy learning. For example, Commission officials observed US trading schemes in action and stated publicly that 'the ETS's "cap and trade" system was inspired by a United States model introduced in the 1990s to curb acid rain' (European Commission 2006: 2). Domestic politics and institutional obstacles also played a role. In the early 1990s, the Commission realized that it would face a difficult if not impossible battle with the member states over a common carbon/energy tax because fiscal instruments require unanimous support in the Council of Ministers. Since the Commission was unlikely to convince all member states to agree to the tax, it began promoting carbon trading. The combination, therefore, of international policy learning and domestic political-institutional constraints highlight the pressures coming from different levels. This change of approach has placed the EU in an international 'leadership' role by making the EU the most important advocate of emissions trading within the Kyoto framework (Wettestad 2005).

As its international role and commitment evolved, the EU began to push for a domestic Europe-wide ETS – an initiative that, crucially, was supported by important economic actors as a new market to complement any future international emissions trading schemes. Despite the costs, the EU moved forward very rapidly in establishing the new instrument (Oberthür and Tänzler 2007). The speed with which this happened is striking for two reasons: (i) the EU lacked previous experience with this market-based mechanism; and (ii) its advocates had to, and did, overcome obstacles within the EU's complex policy-making process quickly and skilfully.

Obstacles to EU climate policy

Despite the EU's apparent success in its multi-level engagements with climate policy, it faces a number of international and domestic political obstacles to more vigorous action on climate policy. Given the multitude of significant veto points during international negotiations and the development of internal policies, this section focuses on key selected international and regional obstacles facing the EU and its flagship ETS, many of which are shared with other aspects of EU climate policy.

International obstacles to EU climate policy

While climate change mitigation is clearly in the interest of all states, the means through which responsive polices will be negotiated and promulgated internationally remain subject to the specific domestic politics of individual states and the variety of public- and private-sector actors involved in these politics. At the international level, obstacles include the need to overcome the conflicting interests of the negotiating parties in the UNFCCC, and obstacles to initiatives promoting global environmental governance (Vogler 2005).

First and foremost, the EU must consider the role played by the Asia-Pacific Partnership on Clean Development and Climate (APP). The APP was launched in January 2006 as a non-treaty agreement and currently includes Australia, Canada, China, India, Japan, Republic of Korea and the USA. APP members account for about half of the world's population, economic output, greenhouse gas emissions and energy consumption. They also produce about 65 per cent of the world's coal, 48 per cent of the world's steel, 37 per cent of the world's aluminium and 61 per cent of the world's cement (APP 2008). The APP's priorities focus on technology-based solutions, and a determination that members should be allowed to set their own goals for reducing emissions individually, with no mandatory enforcement mechanisms. The EU accepts technological solutions as *additional* measures to combat climate change; however, the EU's firm advocacy of binding enforcement mechanisms makes it uncertain how far it will be able to pursue compromise with the APP.

Second, a fully and consistently operable EU ETS will place the EU in a good position to sustain its international leadership role by exploiting first-mover advantages and potential linkages to other emerging emissions trading schemes in countries such as Australia. According to the Commission, 'the ETS is open to linking with compatible greenhouse gas emission trading schemes in other countries that have ratified the Kyoto Protocol. It is foreseen that each side would agree to recognize allowances issued by the other, thereby expanding the market for trading' (European Commission 2005). The Union has also recently confirmed EU ETS participation by three non-EU states: Norway, Iceland and Lichtenstein.

When the EU ETS began operating in a pilot Phase on 1 January 2005, the member states granted emissions permits (allowances) for three years until 2007 to large emitters such as factories and power stations, mainly for free. In April and May 2006, however, the carbon market crashed when the price of permits to emit a tonne of CO_2 plunged 72 per cent to €8.60 in just three weeks. This was precipitated by a series of data releases which showed the EU ETS had a major surplus of allowances caused by member states issuing greater numbers of permits than were required to cover actual emissions in order to protect their energy sectors and trade-exposed industries (Grubb and Neuhoff 2006). In essence, this failure revealed the ever-present tensions between national self-interest, national sovereignty and EU solidarity on climate change. Similar tensions emerged over allocations for the period 2008–2012, although the Commission has taken a stronger stance with the member states, insisting that many governments reduce their national allocations, and is seeking an EU-wide emissions cap from 2013 onwards (Bailey 2007). These experiences nevertheless reveal potential implementation problems that could undermine the EU ETS credibility and the EU's prospects for leadership at the international level.

Similarly, the strategy of linking the EU ETS to other national, regional and international emissions trading schemes (Oberthür 2006; Legge 2007) will have to overcome a number of obstacles related to the technological compatibility, economic/financial viability and political feasibility of linking schemes.

Regional obstacles to the EU emissions trading scheme

The EU also faces internal regional obstacles to the legalities and modalities of the EU ETS. In particular, avoiding another price crash will require continued and robust scrutiny of national allocations. The most important obstacle in this regard may be the way National Allocation Plans (NAPs) are formulated and approved. The NAPs remain a controversial issue among the Union's environmental leaders and laggards, with many member states disagreeing with Commission allocation decisions. In August 2007, Latvia joined Poland, Czech Republic, Slovakia, Hungary and Estonia in taking the Commission to the European Court of Justice over specific emissions calculations and whether the Commission has the right to influence member states' choice of energy supply by imposing national emissions caps. Such legal challenges reflect very real practical (the tendency for member states to seek over-allocations) and political tensions in the development of regional emissions trading schemes (Bailey 2007).

As well as these qualitative obstacles, quantitative obstacles obstruct progress towards the enhanced goals set out in the European Council's Summit in March 2007. Achieving these will require improved performance from all member states and, along with the Commission's proposal to move from the predominant free issue of emissions permits towards up to 80 per cent auctioning, will exacerbate frictions with some industry groups. The EU's 27 members must grapple with different starting points and different abilities to reach these targets (Legge 2007), while also ensuring that aviation emissions are dealt with appropriately.

Important EU member states have already expressed opposition to the Commission's blueprint for a post-2012 climate change regime. For example, French President Nicolas Sarkozy wrote to the President of the Commission stating that:

> Some of the pending proposals are 'neither efficient, fair nor economically sustainable' for France...'European constraints would push industry to shift production to these countries [without similar carbon reduction obligations]. Global emissions would not fall and jobs would disappear from Europe'... French officials have reportedly also consulted their German counterparts on how to react.
>
> (Kubosova 2008a)

The French Government is by no means the only actor to identify possible negative impacts on certain industrial sectors as an obstacle to more vigorous EU climate policies. The Commission's initiatives are expected to affect, to varying degrees, energy-intensive industries such as aluminium, cement, chemicals, fertilizers, pulp and paper and steel. As Kubosova (2008a) notes, 'these industries are expected to have to raise their prices under the more stringent green rules, weakening their position against competitors from other economic superpowers such as the USA or China'.

The EU's climate policies are also likely to face lobbying pressure from labour and other societal groups. Trade unions have already urged the Commission to delay a package of new climate policies 'rather than introduce it without measures designed to soften its "social impact"' (Kubosova 2008a). This package, which includes the review of the EU ETS in preparation for the post-2012 regime, also focuses on other changes necessary to achieve the EU's desired 20 per cent cut in emissions below 1990 levels by 2020 and the expansion in renewable energy articulated in the European Council's decision in March 2007. European labour leaders do acknowledge consultation with EU officials during the review process, but the secretary general of the European Trade Union Confederation has asserted that his organization would like a '"European low-carbon economy adjustment fund" to help workers affected by job losses, as well as a carbon levy on imports to protect Europe's heavy industry from competition from abroad' (Kubosova 2008a).

Leaders of European environmental and development NGOs can be expected to maintain their claims that the Union's initiatives do not go far enough or that they create distorted effects. For example, opposition has already been voiced over the EU's goal of obtaining ten per cent of transport fuels from biofuels by 2020. A group of 17 NGOs – including Oxfam and Friends of the Earth – wrote to the EU's Energy Commissioner in January 2008 asking for tougher standards. Among their concerns were a lack of protection for important ecosystems and water and soil resources, as well as the unintended consequences of increasing food and feed prices, and exacerbating water scarcity, that would negatively impact the world's poor (Kubosova 2008b). The Commission has sought to develop a relatively open decision-making structure to ensure the practicality and acceptability of its climate policies. However, by definition this expands the range of actors that can pressurize the policy process via lobbying.

Political strategies for future EU climate policy

All 27 member states as well as the various EU institutions have their own interests and strategies on climate change, creating an institutional complexity that often confounds efforts to identify a single strategic actor. This section simplifies this complexity by focusing on the political strategies open to the Commission as the main initiator of new EU strategies and overseer of their implementation. Unlike the national polities examined in this book, the Commission does not face direct electoral pressures, but its system of active stakeholder engagement demonstrates that it is certainly not insensitive to outside opinion. However, interactions with public opinion tend to be mediated through the European Council, the various Councils of Ministers and the European Parliament. This presents unique opportunities and constraints in relation to the political strategies available to the Commission to achieve deeper cuts in greenhouse gas emissions.

International strategies

The EU remains a prominent actor in international climate negotiations and, during the UN Climate Change Conference in Bali in December 2007, asserted a bold new position. Many expected the Bali negotiations to focus on a roadmap that would deal with the procedural issues of launching and organizing the post-2012 regime. On the EU's insistence, however, the resulting Convention's objective of preventing dangerous levels of climate change refers to a section of the IPCC's recent Fourth Assessment Report 'which demonstrates that emissions reductions for developed countries in the range of 25–40 per cent below 1990 levels by 2020 are required to limit global warming to 2 degrees above pre-industrial levels' (European Union 2007). This assertion generated significant opposition from some national parties and, at the time, might have seemed an unproductive strategy that could have jeopardized the launch of the negotiations. The insertion of this section in the Convention, however, seems to have vindicated this bold strategy.

Despite this early success, the Union must develop further strategies to garner support among other UNFCCC parties if it hopes to shape the post-2012 system in line with its preferences. In particular, the negotiations will have to address emissions targets for Annex II (developing) countries. Here the EU will have to play a prominent role through its input in the forthcoming review of the Kyoto Protocol, scheduled for completion in December 2008, and new incentives and sustained political pressure will be needed to ensure that developed and developing countries agree to future commitments. This will also require careful tracking of the shifting coalitions among other parties, both developed and developing countries.

As the weight of scientific evidence on climate changes increases, adjustments in government policies among the Annex I countries – especially the APP – may change the nature of international climate politics. The EU must monitor closely, and respond to, these adjustments. Such strategies include intensified public information campaigns in the APP countries and concerted diplomatic efforts targeted at APP members with new governments, in particular Australia and, soon, the USA. The EU may also need to take forward a threat of additional levies on products coming from states that have not ratified the Kyoto Protocol, although it will need to be careful not to contravene World Trade Organization (WTO) trade rules.

To pressure developing countries, the EU may consider more positive strategies, such as linking aid and trade packages to specific emission reduction goals. As the world's largest aid donor and trading bloc, and a significant source and destination of foreign investment, the EU possesses considerable economic leverage to encourage reforms in developing countries. Although the EU's rather tarnished reputation acquired in previous trade negotiations with developing countries may undermine the credibility of this tactic, it might be able to encourage some countries to adopt specific agreements under the post-2012 regime by linking these to Union support for WTO membership. Many non-WTO members (and WTO members) will resist such pressure, but several countries that are not yet full members – for example Algeria, Azerbaijan, Belarus, Ethiopia, Iran, Kazakhstan, Laos, Libya, Sudan, Ukraine, Uzbekistan and Yemen – may be susceptible if the offer is part of an integrated package that covers other policy areas as well.

Much of the EU's international strategy will also depend on how successfully it implements its internal climate change policies. Ensuring effective functioning of the EU ETS, and increasing technological and professional coordination between financial industries involved in the EU ETS, will increase support for linkages to other trading schemes. Setting a normative example at home may, therefore, be an effective strategy for changing ideas and policies abroad.

Regional strategies

At the regional level, the Commission's political strategizing must first and foremost recognize the crucial role and reasoning of the member states in determining the adoption of new climate policies. Member states often challenge EU climate directives not because they are anti-environment but because they are concerned about whether policy decisions are best made in national capitals or Brussels; the extent to which such decisions bind them into further integration; and the implications of ambitious EU policies for their economic competitiveness. The Commission has, of course, faced similar challenges across many policy areas and has developed well-known strategies to cope with them, such as a strong emphasis on scientific evidence in proposals, widespread stakeholder consultation and deliberately tabling overambitious proposals knowing that whatever measures are put forward are likely to be negotiated down in the Council of Ministers (Sbragia 1998; Jordan 2005).

Alongside these standard recipes, specific strategies to promote the EU's internal climate policies must first include provisions to manage cooperation among its enlarged membership of 27 member states. Following the 2004 and 2007 accessions, the EU faces the additional challenge of ensuring robust policy implementation in a number of relatively poor new member states that have strong development needs and ambitions, and a poor track record on monitoring and enforcement. Many of these states also rely on heavily polluting lignite and ageing nuclear power facilities for energy production. It is no coincidence that all the member states that challenged the Commission's decisions on national emissions caps for Phase two of the EU ETS were recent accession countries (Massai 2007). Transitional periods will be required but must be managed carefully to avoid exacerbating divisions between environmental leader and laggard states. Building public support in the new member states will also be problematic due to the low priority of environmental issues in these countries, as indicated by a recent finding that 'more than 62 per cent in the new Member States fear to be without a job and only 3 per cent think that environmental protection is the most pressing problem' (Böhm 2006: 241–2).

A second strategy, which is already being pursued but has significant remaining potential, is to promote more integrated policy-making across policy areas to increase the co-benefits arising from climate policies and, hence, their acceptability to member states and other stakeholders. One example is the linking of climate-related strategies to energy security; another is the use of revenue from EU ETS allowance auctions to support tax cuts or other economic stimuli. The EU's Action Plan on Energy, adopted at the European Council of March 2007, calls for 'concrete actions to achieve a competitive, sustainable and secure energy system' in parallel with greenhouse gas reductions (European Commission 2007: 10). It also sets goals for energy policy linked to energy efficiency for appliances, expansion in renewable energy production, increased use of biofuels and the use of carbon capture and storage. The EU will also to need to ensure a tight focus on sustainable biofuels production in order to meet the concerns of NGOs and others about the potential adverse effects of increased biofuels use on agriculture producers at home and abroad.

Another key component of securing energy security co-benefits will be to formulate an effective foreign policy that addresses its dependence on non-Union (in particular, Russian) energy sources. Because decisions on the structure of energy supply require unanimous support within the Council of Ministers, the EU's ability to intervene on this front is restricted and pursuit of this important (and highly popular) co-benefit may require proposals that link energy policy goals with changes in other single market policies on which decisions can be taken by qualified majority.

A third political strategy needed to meet emissions targets is further broadening of the scope of climate policies, in particular to encompass transport and non-carbon gases. The decision to include aviation in Phase 3 of the EU ETS represents an important step in this direction; however, coverage of other transportation sectors – in particular, shipping and automobiles – will be a contentious but necessary political objective. The Commission's proposal to extend the EU ETS to all greenhouse gases should further enhance the scheme's impact but will also add complicated and contentious new dimensions to the monitoring and enforcement of EU climate policy. Such measures will certainly encounter varying levels of opposition from different member states and stakeholder groups. The Commission will have to build coalitions of support among diverse political and economic actors, taking care to identify the common public- and private-sector interests served by incorporating other sectors and gases into the EU ETS. Given this landscape, the EU must develop strategies supported by financial service providers and other sectors that stand to benefit from emissions trading.

Another tactic for broadening the base of EU climate policy is further expansion of renewable energies. Ensuring the political acceptability of this to the member states will require gradualism in the way targets are increased and differentiation between member states based on their capabilities. The Commission has already taken steps in this direction, setting criteria for determining contributions based on member states' geographical potential to produce energy from different renewable sources and their economic capacity to support investment as indicated by GDP per capita (Goldirova 2007). The political sensitivities involved in adjudicating these criteria will require the Commission to develop a convincing methodology for determining national capabilities that will be acceptable to all member states, or face further cases before the European Court of Justice.

Fourth and finally, further reforms are required to the process used to allocate national emission permits among its member states. National allocations for the EU ETS have become more realistic during Phase two of the scheme – aided by better data from monitoring activities and verified emissions during the trial period – which should reduce, but not necessarily eliminate, the likelihood of inflated claims of need and future market crashes. The political heat can theoretically be taken out of this issue further if the member states accept the Commission's principle of an EU-wide emissions cap. Disputes over the issue of allocations may also be reduced by the Commission's proposals to increase the auctioning of EU ETS permits to 80 per cent (Mandell 2005; Hepburn *et al.* 2006), as this would privilege market forces over political arguments about national need as the mechanism to allocate permits. However, national allocations remain a politically sensitive issue and careful judgement will be required to ensure that the new approach remains sensitive to the differing development needs, energy structures and abatement potentials of different member states.

At first glance, the EU's uniquely complex institutional and sovereignty-sharing arrangements might seem to militate against it being a major actor in climate politics and policy. However, its position as a permanent and authoritative point of national and international interaction also provides it with significant opportunities to influence climate policy at multiple levels, while the Commission's relative distance from direct electoral pressures enables it to develop more ambitious proposals than some of its member states would otherwise contemplate. Despite this, all EU policies remain subject to national scrutiny via the Council of Ministers, so that EU climate policies both transcend and remained strongly tied to national political interests.

In the final analysis, much of the EU's domestic and international credibility in climate policy may hinge on the fortunes of the EU ETS. If the scheme is successful in reducing emissions, it is likely to stimulate further policy and technological innovations as well as enhanced policy learning and diffusion to other regions. A fully functional EU ETS should also create a first-mover advantage in lucrative financial services and set the seal on the EU's reputation as a major player in international and regional climate policy. Conversely, weaknesses in the scheme are likely to be seized upon by certain member states and UNFCCC parties as a justification for the continuation of more conservative climate policies.

References

Andresen, S. and S. Agrawala (2002), 'Leaders, pushers and laggards in the making of the climate regime', *Global Environmental Change* 12, 41–51.

APP (Asia-Pacific Partnership on Clean Development and Climate) (2008), 'Introduction', www.asiapacificpartnership.org/ [13 January 2008].

Bailey, I. (2007), 'Neoliberalism, climate governance and the scalar politics of EU emissions trading', *Area* 39, 431–42.

Betz, R., K. Rogge and J. Schleich (2006), 'EU emissions trading: an early analysis of national allocation plans for 2008–2012', *Climate Policy* 6, 361–94.

Böhm, M. (2006), 'Environmental consequences of enlargement', in Roy, J. and R. Domínguez (eds), *Towards the Completion of Europe*, Coral Gables, Miami: European Union Center, pp. 237–42.

Börzel, T. (2000), 'Why there is no "southern problem". On environmental leaders and laggards in the European Union', *Journal of European Public Policy* 7, 141–62.

Cass, L. (2005), 'Norm entrapment and preference change: the evolution of the European Union position on international emissions trading', *Global Environmental Politics* 5, 38–60.

Christiansen, A. and J. Wettestad (2003), 'The EU as a frontrunner on greenhouse gas emissions trading: how did it happen and will the EU succeed?' *Climate Policy* 3, 3–18.

Damro, C. and P. Luaces Méndez (2003), 'Emissions trading at Kyoto: from EU resistance to Union innovation', *Environmental Politics* 12(2), 71–94.

Delbeke, J. (ed.) (2006), *EU Environmental Law: The EU Greenhouse Gas Emissions Trading Scheme*, EU Energy Law Vol. IV, Leuven, Belgium: Claeys and Casteels.

European Commission (2005), *EU Emissions Trading: An Open Scheme Promoting Global Innovation*, Luxembourg: Office for Official Publications of the European Communities.

European Commission (2006), 'Countering climate change', *EU Insight*, Special EU Advertising Supplement, Delegation of the European Commission to the USA, October 2006.

European Commission (2007), 'EU Action against climate change: leading global action to 2020 and beyond', Luxembourg: Office for Official Publications of the European Communities.

European Commission (2008), 'Proposal for a decision of the European Parliament and of the Council on the effort of member states to reduce their greenhouse gas emissions to meet the Community's greenhouse gas emission reduction commitments up to 2020', Brussels, 23.1.2008 COM(2008) 17 final.

European Union (2007), 'Climate change: EU welcomes agreement to launch formal negotiations on a global climate regime for post-2012', *Press Release*, Bali/Brussels, 15 December, Reference: MEMO/07/588.

Goldirova, R. (2007), 'Focus on renewables: EU plans to boost green energy take shape', www.EUObserver.com [21 December 2007].

Grubb, M. and K. Neuhoff (2006), 'Allocation and competitiveness in the EU emissions trading scheme: policy overview', *Climate Policy* 6, 7–30.

Grubb, M. and F. Yamin (2001), 'Climatic collapse at The Hague: what happened, why, and where to go from here?' *International Affairs* 77, 261–76.

Gupta, J. and M. Grubb (eds) (2000), *Climate Change and European Leadership. A Sustainable Role for Europe?* Dordrecht: Kluwer Academic Publishers.

Gupta, J. and L. Ringius (2001), 'The EU's climate leadership: reconciling ambition and reality', *International Environmental Agreements: Politics, Law and Economics* 1, 281–99.

Hepburn, C., M. Grubb, K. Neuhoff, F. Matthes and M. Tse (2006), 'Auctioning of EU ETS phase II allowances: how and why?' *Climate Policy* 6, 137–60.

Jordan, A. (ed.) (2005), *Environmental Policy in the European Union*, second edition, London: Earthscan.

Jordan, A. and A. Schout (2006), *The Coordination of the European Union: Exploring the Capacities of Networked Governance*, Oxford: Oxford University Press.

Kubosova, L. (2008a), 'Brussels accused of ignoring social concerns in climate bill', www.EUObserver.com [15 January 2008].

Kubosova, L. (2008b), 'EU admits biofuel target problems', www.EUObserver.com [14 January 2008].

Legge, T. (2007), 'An EU outlook on the future of the Kyoto Protocol', *The International Spectator* 42, 81–93.

Lenschow, A. (2005), 'Environmental policy', in Wallace, H., W. Wallace and M. Pollack (eds), *Policy-making in the European Union*, fifth edition, Oxford: Oxford University Press, pp. 305–28.

Lisowski, M. (2002), 'Playing the two-level game: US President Bush's decision to repudiate the Kyoto Protocol', *Environmental Politics* 11, 101–19.

Mandell, S. (2005), 'The choice of multiple or single auctions in emissions trading', *Climate Policy* 5, 97–107.

Massai, L. (2007), 'Climate change policy and the enlargement of the EU', in Harris, P. (ed.), *Europe and Global Climate Change*, Cheltenham: Edward Elgar, pp. 305–21.

Mazey, S. and J. Richardson (1992), 'Environmental groups and the EC: challenges and opportunities', *Environmental Politics* 1(4), 109–28.

McCormick, J. (2001), *Environmental Policy in the European Union*, Basingstoke: Palgrave.

Michaelowa, A. (1998), 'Impact of interest groups on EU climate policy', *European Environment* 8, 152–60.

Oberthür, S. (2006), 'The climate change regime: interactions with ICAO, IMO, and the EU Burden-Sharing Agreement', in Oberthür, S. and T. Gehring (eds), *Institutional Interaction in Global Environmental Governance*, Cambridge MA: MIT Press, pp. 53–77.

Oberthür, S. and D. Tänzler (2007), 'Climate policy in the EU: international regimes and policy diffusion', in Harris, P. (ed.), *Europe and Global Climate Change*, Cheltenham: Edward Elgar, pp. 255–78.

Sbragia, A. (1998), 'Institution-building from below and above: the European Community in global environmental politics', in Sandholtz, W. and A. Stone Sweet (eds), *European Integration and Supranational Governance*, Oxford: Oxford University Press, pp. 283–303.

Skodvin, T. and S. Andresen (2006), 'Leadership revisited', *Global Environmental Politics* 6, 13–27.

Steurer, R. (2003), 'The US's retreat from the Kyoto Protocol: an account of a policy change and its implications for future climate policy', *European Environment* 13, 344–60.

Vogler, J. (2005), 'Europe and global environmental governance', *International Affairs* 81, 835–50.

Watanabe, R. and G. Robinson (2005), 'The European Union Emissions Trading Scheme (EU ETS)', *Climate Policy* 5, 10–14.

Wettestad, J. (2005), 'The making of the 2003 EU Emissions Trading Directive: an ultra-quick process due to entrepreneurial proficiency?' *Global Environmental Politics* 5, 1–23.

Winne, S., A. Haxeltine, W. Kersten and M. Berk (2005), 'Towards a long-term European strategy on climate change policy', *Climate Policy* 5, 244–50.

Woerdman, E. (2004), 'Path-dependent climate policy: the history and future of emissions trading in Europe', *European Environment* 14, 261–75.

Zito, A. (2005), 'The European Union as an environmental leader in a global environment', *Globalizations* 2, 363–75.

5

Federal Climate Politics in the United States: Polarization and Paralysis

Paul R. Brewer and Andrew Pease

Introduction

The debate over global climate change came to movie theatres in the United States with the 2006 release of *An Inconvenient Truth*, a documentary presentation of evidence that human-produced greenhouse gases are contributing to such change. The film starred Al Gore, who served as vice president of the United States from 1993 to 2001 and ran unsuccessfully for the presidency in 2000 as the nominee of the Democratic Party. *An Inconvenient Truth* was a commercial and critical success, earning over $23 million in the United States and winning an Academy Award for Best Documentary (Internet Movie Database 2007). In broader terms, the documentary not only generated news media coverage and helped fuel public debate; it also appeared to resonate with public opinion. Polls showed that majorities of the public believed that global climate change was occurring, that human activity was contributing to it, that it was a serious problem, and that government should act to address it (see, for example, Bowman 2007; Polling Report 2007a). Nevertheless, the release of *An Inconvenient Truth* failed to generate public policy shifts on the part of the US government. Indeed efforts to address climate change at the federal level have met with little success over the past two decades.

US policy efforts to address climate change

The first major effort by the US government to address climate change occurred in 1991 when President George H.W. Bush unveiled his Action Agenda on Climate Change. The lack of binding emissions reduction targets in the Agenda reflected the administration's ambivalence on climate change. On one hand, the release of the Intergovernmental

Panel on Climate Change's (IPCC) First Assessment Report in 1990, and the growing international consensus regarding climate change, made it impossible for Bush, a self-professed 'environmental president', to ignore the issue completely. On the other hand, the Bush presidency was still reeling from a broken promise not to raise taxes. Despite a 1991 report issued by the National Academy of Sciences (1991: 73) stating that 'the United States could reduce or offset its greenhouse gas emissions by between 10 and 40 per cent of 1990 levels at low cost, or at some net savings, if proper policies are implemented', conservative Republicans in Congress were convinced that any binding resolution to reduce greenhouse gas emissions would entail a steep rise in energy taxes.

With the 1992 presidential race looming, Bush could not afford to enter into an agreement that would further antagonize the conservative bloc of his party. During the negotiations leading up to what would eventually become the United Nations Framework Convention on Climate Change (UNFCCC), the Bush camp refused to commit to a binding agreement to reduce carbon dioxide emissions by a specific date. The final version of the UNFCCC, which the Senate ratified on 7 October 1992, contained only 'non-binding' voluntary reduction goals, requiring that all signatory nations commit to achieving 'stabilization of greenhouse gas concentrations in the atmosphere at a level that would prevent dangerous anthropogenic interference with the climate system' and that they prepare a 'national action plan to address emissions of greenhouse gases' (Parker and Blodgett 2007: 1).

In accordance with the steps outlined in the UNFCCC, Bush released the National Action Plan for Global Climate Change in December 1992. The Plan included numerous carbon dioxide reduction initiatives, highlighted by Environmental Protection Agency programmes such as Energy Star, which promoted energy-efficient consumer products, and Green Lights, which encouraged US corporations to install energy-efficient lighting. The voluntary (as opposed to regulatory) nature of the various initiatives included in the Plan further underscored the limits of the Bush administration's approach to climate change. The Bush policy, dubbed 'no regrets', did not go beyond advocating programmes that were justifiable for reasons other than reducing greenhouse gas emissions, such as energy conservation or pollution control, and that could be implemented at no cost to taxpayers (Yacobucci and Parker 2006). The concomitant emissions reductions were viewed as an ancillary bonus, but the perceived scientific uncertainty regarding climate change was deemed to be too great to justify taking substantive actions with the sole purpose of reducing greenhouse gas emissions (Parker and Blodgett 1999).

Bill Clinton's victory in the 1992 presidential election buoyed hopes for the implementation of more aggressive federal policies regarding climate change. The Clinton administration's legacy regarding climate change, however, was largely one of failed efforts and unfulfilled promise. In 1993 Clinton attempted to implement a British Thermal Unit (BTU) tax that 'would have taxed virtually all forms of fossil fuel energy in the US' (Fisher 2004: 122). Despite the control of both the House and the Senate by his fellow Democrats, Clinton was unable to muster enough votes to push the BTU tax through Congress. Fierce opposition from the business lobby convinced a number of Democratic senators to cross party lines and oppose the levy. Clinton ultimately backed away from his proposal following a closed-door strategy session with key Senate Democrats in which he was allegedly told that the odds of the Senate passing the BTU tax in the form that he proposed were 'extremely gloomy' (Rosenbaum 1993: B1). In many respects the failed BTU tax may have been the Clinton administration's best opportunity for successfully implementing a meaningful climate change policy initiative. The 1994 midterm elections, which ushered in a Republican majority in Congress, effectively doomed the administration's future efforts to pass binding climate change legislation.

The growing partisan divide over climate change policy, already evident in the debates surrounding the failed BTU tax, was exacerbated in 1995 by the Clinton administration's decision to endorse the Berlin Mandate despite fervent objections from congressional Republicans. Given that the majority of the UNFCCC's signatory nations had failed to reduce their levels of greenhouse gas emissions on a voluntary basis, the Berlin Mandate was designed to provide a roadmap for a future international agreement (eventually, the Kyoto Protocol) that would include legally binding and enforceable targets (Fisher 2004). The Berlin Mandate's provision to exempt developing nations from binding commitments was the chief cause of discord, as many in Congress believed that its inclusion would place the United States at an economic disadvantage in relation to exempt nations such as India and China. Nevertheless, Clinton signed the 1996 Geneva Declaration, which stipulated that the next round of negotiations, to be held the following year in Kyoto, would produce an agreement that included legally binding emissions targets which, because of the Berlin Mandate, would apply only to Annex I ('developed') nations (Fisher 2004).

In a preemptive strike designed to assert the parameters under which it would approve an agreement emerging from the Kyoto negotiations, the Senate unanimously passed Senate Resolution 98, better known as the Byrd–Hagel Resolution (Fisher 2004), which stated that the United States should not sign any international agreement on greenhouse gas reduction that exempted developing nations. In addition, an agreement would be unacceptable if it could 'result in serious harm to the economy of the United States' (Senate Report 105–54 1997: 24).

As expected, the final version of the Kyoto Protocol did not meet the guidelines set forth in the Byrd–Hagel Resolution. Nevertheless, Clinton chose to make the United States a signatory nation, calling the Protocol a work in progress (Fisher 2004). At the same time, he stated that he would not submit it to the Senate for ratification until at least some non-Annex I nations committed to binding emissions targets (Parker and Blodgett 1999). The key non-Annex I nations did not make this commitment; consequently the Kyoto Protocol languished, signed but not ratified, for the duration of the Clinton administration.

For practical purposes, the Clinton administration's climate change policy did not deviate significantly from the previous Bush administration's policy of 'no regrets'. Clinton's Climate Action Plans outlined a host of voluntary measures, such as a 'Golden Carrot' programme to induce improvements in the efficiency of industrial equipment, a renewable energy consortium, a programme to encourage employers to replace parking subsidies with cash incentives for shared commuting and a programme to promote more efficient nitrogen fertilizer use (Yacobucci and Parker 2006). Although the specific policy initiatives had changed, the net result did not. Despite the various voluntary initiatives enacted during the 1990s, emissions continued to rise. Just prior to the Kyoto negotiations in 1997, carbon dioxide emissions were already 12.9 per cent above 1990 levels (Fisher 2004).

In 2001 Clinton was succeeded by a Republican, George W. Bush. On 11 June of that year, the new president formally rejected the Kyoto Protocol, saying that it was 'fatally flawed in fundamental ways' (Yacobucci and Parker 2006: 5–6). His arguments echoed the Congressional concerns codified in the Byrd–Hagel Resolution, namely that the Kyoto Protocol exempted China and other large developing nations and that substantial reductions in greenhouse gas emissions would entail substantial economic costs. Instead of mandatory emissions reductions, Bush advocated voluntary measures aimed at reducing greenhouse gas intensity, that is, the ratio of greenhouse gas emissions to economic output (Fletcher and Parker 2007). Bush's 2002 Climate Action Report followed the same 'no regrets' policy course charted by the previous two administrations. Of the 50 initiatives summarized in the report, only six were described as regulatory (as opposed to voluntary), and the programmes described as regulatory were implemented to meet other objectives while producing concomitant emissions reductions (Yacobucci and Parker 2006).

In the absence of executive leadership on the issue, some members of Congress attempted to fill the void. As of 17 July 2007, members of the 110th Congress had introduced 54 bills – 28 in the Senate and 26 in the House – that directly addressed climate change (Ramseur and Yacobucci 2007). Developments also occurred on the judicial front. On 2 April 2007, the US Supreme Court determined in *Massachusetts* v. *EPA* that the Clean Air Act authorizes the EPA to regulate emissions from new motor vehicles on the basis of their possible climate change impacts.

Although the Bush administration has consistently opposed entering into any international agreement that includes binding emissions targets, it has been more active recently in attempting to engage foreign nations – especially developing economic powers such as India and China – in cooperative efforts to reduce greenhouse gas intensity through voluntary measures. In July 2005 the United States announced the formation of the Asia–Pacific Partnership on Clean Development and Climate (APP). This six-nation partnership, consisting of China, India, South Korea, Japan, Australia and the United States, focused on developing new technologies to reduce greenhouse gas intensity (Fletcher and Parker 2007).

In September 2007, Bush also invited representatives of 15 of the world's largest economies to the White House for a two-day conference on global climate change. In his address to the gathered delegates, Bush called on 'all the world's largest producers of greenhouse gas emissions, including developed and developing nations', to come together and 'set a long-term goal for reducing' greenhouse emissions (White House 2007). 'By setting this goal, we acknowledge there is a problem, and by setting this goal, we commit ourselves to doing something about it', said Bush. This shift in rhetoric, however, was not matched by substantive policy change.

Political obstacles to US action on climate change

Before looking at specific obstacles to US policy action on climate change, it is important to note two structural features of the national government that serve to make any sort of policy action difficult. The first is the system of checks and balances that the US Constitution imposes on the policy-making process. In order to be enacted as law, a piece of legislation must win majority approval in both chambers of the US Congress – the House of Representatives and the Senate – and then be signed by the president. In practice, a minority of 40 (out of 100) US senators can block legislation through a procedural manoeuvre known as the filibuster. A presidential veto of a bill can only be overridden by a two-thirds vote in both chambers. In addition, treaties negotiated by the president must be ratified by a two-thirds vote of the Senate. These checks and balances provide multiple routes for blocking policy change.

The second structural impediment to policy-making is the potential for – and frequent presence of – divided party control of government. Two parties, the Democrats and the Republicans, dominate national politics. Throughout most of the past two decades these parties have split control of the presidency, the Senate and the House. When Republican George H.W. Bush won the presidency in 1988 he faced a Congress in which the Democrats controlled both chambers. Although Democrat Bill Clinton succeeded him in 1992, unified control of government lasted only until the Republican takeover of Congress in 1994. Republican George W. Bush enjoyed a brief period of unified control in 2001 and a longer one from 2002 to 2006, but the Democrats won control of both chambers in 2006. In times of divided government, policy changes are particularly difficult to enact without bipartisan support.

These structural obstacles might not matter so much if there were a political consensus in favour of policy action on climate change. No such consensus exists in the United States, however. Instead, an anti-environmental movement has been active in opposing new efforts to address climate change through public policy. This movement combines powerful economic interests – including the automobile, coal and oil industries – with conservative organizations and Republican politicians (see, for example, McCright and Dunlap 2000; 2003). The fossil fuel industries have typically seen policy proposals on climate change as moves to impose costs upon them and threaten their profits. Conservative organizations and Republican politicians, in turn, have often held extensive connections to, and drawn substantial support from, these industries.

The various actors who make up the anti-environmental movement have expended considerable energy and money in promoting scepticism about scientific evidence regarding climate change and publicizing the potential economic costs of policy efforts to address climate change.

For example, on 28 July 2003 Senator James Inhofe, a conservative Republican, delivered a speech to his colleagues in which he argued that 'the claim that global warming is caused by man-made emissions is simply untrue and not based on sound science'; that 'CO_2 does not cause catastrophic disasters – actually it would be beneficial to our environment and our economy'; that 'Kyoto would impose huge costs on Americans, especially the poor'; and that 'proponents [of Kyoto] favour handicapping the American economy through carbon taxes and more regulations'. He concluded by suggesting that 'man-made global warming is the greatest hoax ever perpetrated on the American people' (Inhofe 2003).

In a pair of studies, McCright and Dunlap examined how conservative think tanks worked to promote such claims from 1990 to 1997. Their first study showed that the websites of these think tanks emphasized uncertainty in the evidence for climate change while emphasizing certainty in the costs imposed by policies intended to combat climate change (McCright and Dunlap 2000). A follow-up study showed how conservative think tanks and climate change sceptics affiliated with fossil fuel industries exploited the 'political opportunity structure' produced by the 1994 Republican takeover of Congress to undermine support for the Kyoto Protocol (McCright and Dunlap 2000: 360). Since 1997 the anti-environmental movement has continued its efforts along these lines. In 2007, for example, the American Enterprise Institute – a conservative think tank funded in part by oil company Exxon Mobil – offered a $10,000 reward for 'for articles [emphasizing] the shortcomings of a report from the UN's Intergovernmental Panel on Climate Change' (Sample 2007).

The activities of the anti-environmental movement help to account for why the information about climate change available to the US public has not reflected the discourse among scientists. The claim that human activity has, and is, influencing climate change dominates scientific discourse. In a synthesis that accompanied its last assessment report, the IPCC – which consists of more than 2000 of the world's leading atmospheric scientists – concluded that '[t]here is new and stronger evidence that most of the warming observed over the last 50 years is attributable to human activities' (2001: 5). An analysis of ten years' worth of peer-reviewed journal articles failed to find one dissenting view on the existence of anthropogenic climate change (Orestes 2004). In the 2 May 2006 executive summary of its first synthesis report, the US Climate Change Science Program (USCCSP), which President George W. Bush had launched in February 2002 to study the causes and potential impacts of climate change, observed that research shows 'clear evidence of human influences on the climate system' (Wigley *et al.* 2006: 2).

News media coverage in the United States has stood in sharp contrast to this consensus. Boykoff and Boykoff (2004) found that from 1988 and 2002 newspaper depictions of scientific discourse diverged significantly from the overwhelming scientific consensus. Looking at a random sample of 636 newspaper articles dealing with climate change, the authors found that 'balanced' accounts comprised slightly more than half of the articles and that almost all (95 per cent) of the articles made at least some mention of dissenting viewpoints. Boykoff (2008) found a similar pattern in broadcast and cable television news coverage: 70 per cent of all segments provided 'balanced' coverage of anthropogenic climate change. Both studies concluded that reporters' adherence to the journalistic norm of balance contributed to the gulf between scientific discourse and popular discourse.

The false balance created by giving equal weight to climate change sceptics may have shaped public understandings (or misunderstandings) of the subject. Research indicates that among citizens with limited cognitive skills, increased exposure to the views of these sceptics produced greater scepticism regarding the existence of climate change (Krosnick *et al.* 2000). Public misconceptions about scientific consensus, in turn, may have undermined popular support for actions to combat climate change. A 2005 survey conducted by the Program on International Policy Attitudes (PIPA) revealed a strong relationship between belief that there was a scientific consensus on climate change and belief that high-cost steps needed to be taken to address the issue. Only 17 per cent of those who believed that scientists were divided on the existence of global warming favoured high-cost steps, compared with 51 per cent of those who perceived a scientific consensus (PIPA 2005).

Elite cues also appear to have shaped public opinion about policy efforts to address climate change. Drawing on two national surveys, one conducted before President Clinton's 1997 push to build support for the Kyoto Protocol and another conducted shortly afterward, Krosnick and his colleagues (2000) found that the resulting debate about climate change produced little if any change in public opinion as a whole. This aggregate stability, however, masked substantial movement among subgroups of the population, with 'Democratic citizens mov[ing] toward the administration's point of view at the same time that Republican citizens moved away' (Krosnick *et al.* 2000: 253). Thus the effect of Clinton's campaign on behalf of Kyoto was to produce further divisions along party lines among the public.

In addition to being divided on the subject of climate change, the American public was relatively apathetic towards it. Bord *et al.* concluded in the late 1990s that 'global warming is not a salient problem for most Americans' and that 'concern for global warming is greater in most other countries' (1998: 84). Almost a decade later the Pew Research Center for the People and the Press (PRC 2007) reached a similar conclusion:

> A survey last year by the Pew Global Attitudes Project showed that the public's relatively low level of concern about global warming sets the US apart from other countries. That survey found that only 19 per cent of Americans who had heard of global warming expressed a great deal of personal concern about the issue. Among the 15 countries surveyed, only the Chinese expressed a comparably low level of concern (20 per cent).

Bowman (2007) offers a range of explanations for why climate change ranked low on the public's agenda, including perceptions that climate change is a 'problem for the future', competition from such pressing issues as the war in Iraq and health care, and an absence of 'tangible manifestations of global warming' for most Americans.

The political landscape in 2007: ongoing polarization

Zaller's (1992) theory of elite signals and public opinion provides a useful framework for analyzing the national political landscape on climate change in 2007. According to Zaller, citizens rely on cues from political elites in forming their opinions about issues. Specifically, citizens judge the credibility of information in public debate by the extent to which it comes from sources that occupy their side of a given belief divide. For example, Republican citizens should be more likely than Democratic citizens to regard Republican leaders as credible, whereas Democratic citizens should be more likely than Republican citizens to regard Democratic leaders as credible.

At the same time, Zaller's model posits that not all people will be equally likely to judge sources on the basis of such belief compatibility. Instead, the politically attentive should be more likely than their less attentive peers to possess the 'contextual information' necessary to do so (Zaller 1992: 42). Given that political attentiveness is closely tied to education, the impact of elite signals on public opinion may vary with level of education. The nature of such conditional relationships may depend, in turn, on the nature of the elite cues available to the public. One potential pattern is the 'polarization effect' (Zaller 1992: 100–2), in which diverging messages from Republican and Democratic elites produce an opinion gap between Republican and Democratic citizens that widens with greater political attentiveness (and, by extension, education).

As of 2007, Republican and Democratic leaders continued to exhibit clear partisan divisions on climate change and policy efforts to address it. One source of evidence for such divisions is the February 2007 Congressional Insiders Poll sponsored by the *National Journal*. Among the 41 Democratic members of Congress polled, 39 agreed that 'it's been proven beyond a reasonable doubt that the Earth is warming because of man-made problems'. In contrast, only four of the 31 Republican members of Congress polled agreed with this statement. Similar gaps emerged on a series of questions about policies designed to address climate change (see Table 5.1).

Likewise, the contenders for the 2008 presidential nominations diverged along party lines on policy efforts to address climate change.

Table 5.1 Views of Congressional Democrats and Republicans on climate change

	Democrats (%)	**Republicans** (%)
Do you think it's been proven beyond a reasonable doubt that the Earth is warming because of man-made problems?	95	13
Which of these actions to reduce global warming could you possibly support?		
... mandatory limits on CO_2 emissions	88	19
... increased spending on alternative fuels	95	71
... higher fuel efficiency standards for automobiles	90	45
... a 'cap and trade' CO_2 emissions reduction program	83	42
... a carbon tax	50	6
... greater reliance on nuclear energy	58	90

Note: For Democrats, $N=41$ (first question) or 40 (other questions); for Republicans, $N=31$.
Source: Congressional Insiders Poll, *National Journal* (2007).

Of the four leading candidates for the Democratic nomination – Senator Hillary Clinton, Senator Barrack Obama, former Senator John Edwards and Governor Bill Richardson – all favoured a cap and trade system that would reduce carbon emissions by at least 80 per cent by 2050, as well as increases in automobile fuel efficiency standards. Of the four leading candidates for the Republican nomination – former Mayor Rudy Giuliani, Senator John McCain, former Governor Mitt Romney and former Senator Fred Thompson – only McCain supported a cap and trade system and increased automobile fuel efficiency standards. Neither Giuliani nor Thompson took a clear position on either proposal in 2007, while Romney supported the cap and trade system only as part of an international cap and did not support increased fuel efficiency standards as a stand-alone measure (Moore 2007).

A January 2007 survey sponsored by the PRC, in turn, provides data regarding public opinion about climate change and policy efforts to address it (see Table 5.2). A clear majority of the respondents (77 per cent) believed that 'the earth is getting warmer', but slightly less than half of the full sample (47 per cent) believed that this was 'mostly because of human activity such as burning fossil fuels' – a figure that was unchanged from July 2006 (PRC 2007). Democratic respondents were more likely than Republican respondents to believe in anthropogenic climate change. In addition, the partisan gap among respondents grew

Table 5.2 US public opinion on climate change, by education and party

	Not college graduate			College graduate		
	Dem.	**Ind.**	**Rep.**	**Dem.**	**Ind.**	**Rep.**
Believe that the earth is getting warmer mostly because of human activity such as burning fossil fuels	52%	49%	32%	75%	43%	23%
Mean rating of how serious a problem global warming is (0 = not a problem, 1 = not too serious, 2 = somewhat serious and 3 = very serious)	2.36	2.22	1.78	2.66	2.21	1.58
Think global warming is a problem that requires immediate government action	69%	69%	44%	88%	66%	32%

Note: N = 1708. 'Dem.' = Democrat; 'Ind.' = Independent; 'Rep.' = Republican.
Source: Pew Research Center for the People and the Press (2007).

as education increased. This pattern suggests that partisan identifiers among the public followed the cues of party leaders.

The results of the Pew survey provide further evidence that climate change ranked relatively low on the public agenda. Only 38 per cent of the respondents rated 'dealing with global warming' as a 'top priority... for President Bush and Congress this year', placing this issue below many others about which the survey asked. Not even a majority of Democrats rated climate change as a top priority (48 per cent did so, versus 40 per cent of independents and 23 per cent of Republicans).

In response to another question, however, 45 per cent of the respondents said that 'global warming' was a 'very serious' problem and another 32 per cent said that it was a 'somewhat serious problem', compared with just 20 per cent who said that it was 'not too serious' or 'not' a problem. Consistent with the partisan pattern in public opinion, Democrats were more likely to say that climate change was a serious problem than were Republicans. Furthermore, the partisan gap on this question was greatest among the most educated respondents.

Yet another example of such polarization came from a question that directly addressed policy action on climate change. Among the sample as a whole, a majority of respondents said that 'global warming is a problem that requires immediate government action' (55 per cent, compared with 31 per cent who said that it did not). Following the now-familiar pattern, Democrats were more likely to endorse immediate government action than were Republicans, with the most educated partisans also being the most likely to side with their party's leaders.

Building a consensus for US policy action on climate change

Despite the efforts of some politicians and activists, the US government has done little to address climate change through anything beyond limited and voluntary programmes. In no small part, this lack of tangible action reflects the endeavours of a powerful anti-environmental movement. By promoting scepticism about climate change, this movement succeeded in constructing a public discourse that failed to reflect the scientific consensus. For their part, both politicians and attentive citizens have been polarized along party lines on climate change – a state of affairs that has made consensus on policy efforts difficult to attain. The broader public, in turn, has been relatively apathetic towards the issue.

In light of the partisan divisions that continued to rule the political landscape as of 2007, the failure of *An Inconvenient Truth* to spark immediate policy change comes as no surprise: the people most likely to respond to Gore's message, namely educated Democrats, were also the ones most likely to agree with that message in the first place. Nor is the 2008 presidential election likely to spur new policy action, at least on its own. If any of the leading Republican candidates other than McCain wins, the new president will probably seek to maintain the status quo on government efforts to address climate change. Even if a Democrat were to win the presidency, enjoy unified party control of Congress and choose to spend political capital on climate change – an issue, again, that ranks relatively low on the public agenda – the Republicans could stop new legislation as long as they continued to hold enough seats in the Senate to sustain a filibuster. Furthermore, Democratic legislators representing districts dependent on fossil fuel industries could face pressures to go against their party on climate change.

All of this suggests that a politically feasible strategy for addressing climate change in the United States must involve steps to produce consensus across party lines. Specifically, efforts at enacting policy change must win substantial support from not only Democrats but also Republicans.

A top-down strategy: elite leadership

One route for building such consensus would be a top-down path: prominent Republicans could go against the mainstream of their party by endorsing more aggressive policy efforts to address climate change. In doing so, they might lead some Republican citizens towards greater support for these efforts. To date, the most visible Republican to challenge the party line on climate change has been California Governor Arnold Schwarzenegger. In 2006 he signed a state law that mandated reductions in greenhouse gas emissions. The following year, he pledged to sue the US Environmental Protection Agency, which had stalled California's plan, in order to force it to grant permission for the state to impose stricter air pollution standards than those set by the federal government (Egelko 2007). That same year he joined with the governors of five other Western states and two Canadian provinces in launching the Western Climate Initiative, a regional agreement on emissions reductions and a carbon-trading system (Roosevelt 2007a). He also joined with outgoing British Prime Minister Tony Blair in calling for world action on climate change (Jordan 2007) and criticized the federal government for failing to show leadership on the issue (Williams 2007).

To be sure, Schwarzenegger's support for policy efforts has its limits. In October 2007, for example, he vetoed bills that would have mandated low-carbon automobile fuel and energy-efficient buildings (Roosevelt 2007b). Moreover, he is a relatively moderate Republican governor of a traditionally Democratic state, making him an anomaly in the national context. Even so, Schwarzenegger is a prominent figure among Republicans, and his public endorsement of mandatory action may send the sort of message that could eventually shift opinions within his party. He has explicitly challenged the partisan divide on the issue: 'There is no Democratic planet Earth. There is no Republican planet Earth. There's just a planet Earth, and we all have a responsibility to take care of it' (Williams 2007: A12).

The political landscape may also provide an opportunity to build elite consensus around a policy approach that encourages the development of nuclear energy as an alternative to fossil fuels. The party roles are reversed here, with Republican elites offering greater support than do Democratic elites. As a case in point, Republican respondents in the *National Journal*'s Congressional Insiders Poll were more likely to favour 'greater reliance on nuclear energy' than were Democrat respondents (see Table 5.1). Still, a majority of the Democrats also favoured greater reliance on nuclear energy. Three of the four leading candidates for the 2008 Democratic presidential nomination also favoured nuclear power (Clinton, Obama and Richardson; only Edwards was opposed), while all four leading Republican candidates favoured it (Moore 2007). The primary obstacle to this solution may be opposition from the broader public. Polls reveal widespread scepticism about nuclear energy (Polling Report 2007b), but an elite consensus in favour of developing nuclear alternatives might shift public opinion.

A bottom-up strategy: grassroots reframing and coalition-building

The top-down path for producing bipartisan consensus on policies to address climate change could be complemented by a bottom-up path involving grassroots activism. In particular, the 'crunchy conservative' movement offers a potential counterbalance to the anti-environmental movement. As defined by Rod Dreher, this vision of conservative politics espouses traditional religious values and the free market but rejects 'the consumerist and individualist mainstream of American life' while casting conservation as a 'moral and patriotic issue' (Dreher 2006: 12, 161). 'Conservatives pride themselves on being hard-headed realists, but our lack of serious concern about the environment does not match reality', writes Dreher (2006: 165).

At present this stance is a minority view among conservatives in the United States, whereas the anti-environmental position dominates conservative politics. Dreher (2006: 166), however, suggests a series of ways in which advocates of policy action on climate change could frame such action as being consistent with conservative concerns and principles:

> The problem is that most of us think about global warming, the depletion of fisheries, and the eradication of the rain forests as 'environmental problems'. In truth, they are far more than that. They are economic problems. They are national security problems. They are 'family values' problems. They are religious problems.

A wide body of public opinion research suggests that activists can shape public opinion by disseminating frames that resonate with citizens' core values, particularly when those frames come from credible sources (see, for example, Nelson *et al.* 1997; Druckman 2001). Thus, a reframing of climate change by pro-policy action conservative activists might facilitate political shifts at the grassroots level.

Evangelical Christians could be a particularly receptive audience to such framing efforts. Indeed, efforts at forging alliances between environmental organizations such as the Sierra Club and evangelical activists have already produced some successes (Eilperin 2007). In 2006, 86 evangelical Christian leaders – including the presidents of 39 evangelical colleges – launched an 'Evangelical Climate Initiative' that called for the federal government to require emissions reductions, and funded television advertisements framing efforts to stop climate change as part of human stewardship of God's creation (Goodstein 2006). Some evangelical Christians also challenged the use of fuel-inefficient sport utility vehicles by asking: 'What would Jesus drive?' These efforts met with a backlash from other evangelical leaders, however, with many prominent figures in the movement signing a 2006 letter that stated, 'Global warming is not a consensus issue' (Goodstein 2006: A12).

Challenging opponents, setting the agenda and seizing key moments

Of course the anti-environmental movement is unlikely to fade away anytime soon. It will undoubtedly react to attempts to forge bipartisan consensus by placing pressure on Republican politicians to maintain the party line and continuing its efforts to promote scepticism about climate change. Guided by its norm of objectivity, the US news media may continue to abet the latter strategy by providing 'balanced' coverage of climate change. As a result, those who seek to promote further policy action must work to challenge both the anti-environmental movement and news media practices. A recent analysis conducted by Boykoff (2007) suggests progress on this front. Looking at newspaper coverage from 2003 to 2006, he found that coverage of climate change in US newspapers shifted substantially away from the 'balanced' pattern described above and towards a reflection of the scientific consensus.

Advocates of more aggressive policy action must also work to raise the issue of climate change on the broader public agenda. *An Inconvenient Truth* may have failed to do so by itself, but wider opinion leadership efforts by credible politicians and activists have the potential to foster public concern. Not all arguments for the importance of climate change may resonate equally with the public, however. For example, one study found that broad-based arguments about the consequences of climate change for rising sea levels, food shortages, water shortages and animal extinctions did more to motivate concern than did arguments about consequences at the personal level (Krosnick *et al.* 2000). Thus a key task is to find specific messages and frames that promote public attention and concern.

A related task is to seek out and take advantage of what Boykoff (2007: 474) calls 'critical discourse moments' – that is, events that challenge the frames in public debate and thus provide openings to reframe that debate. Just as the anti-environmental movement has exploited political opportunity structures, so too must advocates of policy proposals to address climate change. Boykoff's (2007) analysis suggests room for hope on this score. In particular, he points to a confluence of political events (including Bush's recognition of anthropogenic climate change and Schwarzenegger's policy actions), scientific activities (including the leaking of drafts of the USCCSP's report) and ecological events (including the devastation wrought on the Gulf Coast by Hurricane Katrina in 2005) that created conditions conducive to a shift in media coverage of climate change. As the political, scientific and ecological contexts evolve, they may yield new opportunities to raise climate change on the agenda, move public opinion and, ultimately, shape policy.

References

Bord, R. J., A. Fisher and R. E. O'Connor (1998), 'Public perceptions of global warming: United States and international perspectives', *Climate Research* 11, 75–84.

Bowman, K. (2007), *How Hot is Global Warming? A Review of the Polls*, www.aei.org/publications/pubID.26519/pub_detail.asp [2 November 2007].

Boykoff, M. T. (2007), 'Flogging a dead norm? Newspaper coverage of anthropogenic climate change in the United States and the United Kingdom from 2003 to 2006', *Area* 39, 470–81.

Boykoff, M. T. (2008), 'Lost in translation? United States television news coverage of anthropogenic climate change, 1995–2004', *Climatic Change* 86, 1–11.

Boykoff, M. T. and J. M. Boykoff (2004), 'Balance as bias: global warming and the US prestige press', *Global Environmental Change* 14, 125–36.

Congressional Insiders Poll (2007), *National Journal*, syndication.nationaljournal.com/images/203Insiderspoll_NJlogo.pdf [31 October 2007].

Dreher, R. (2006), *Crunch Cons: The New Conservative Counterculture and its Return to Roots*, New York: Three Rivers Press.

Druckman, J. N. (2001), 'On the limits of framing effects: who can frame?' *Journal of Politics* 63, 1041–66.

Egelko, B. (2007), 'Governor postpones suing EPA over vehicle emission standards', *San Francisco Chronicle*, 24 October, B4.

Eilperin, J. (2007), 'Warming draws evangelicals into environmentalist fold', *Washington Post*, 8 August, A1.

Fisher, D. (2004), *National Governance and the Global Climate Change Regime*, Lanham, MD: Rowman and Littlefield Publishers.

Fletcher, S. R. and L. Parker (2007), *Climate Change: the Kyoto Protocol and International Actions*, www.ncseonline.org/NLE/CRSreports/07Jul/RL33826.pdf [1 November 2007].

Goodstein, L. (2006), 'Evangelical leaders join global warming initiative', *New York Times*, 8 February, A12.

Internet Movie Database (2007), *Box Office Business for an Inconvenient Truth*, www.imdb.com/title/tt0497116/business [2 November. 2007].

Inhofe, J. M. (2003), *The Science of Climate Change*, inhofe.senate.gov/pressreleases/climate.htm [2 November 2007].

IPCC (Intergovernmental Panel on Climate Change) (2001), *Climate Change 2001 Synthesis Report*, www.ipcc.ch/pub/syreng.htm [2 November 2007].

Jordan, M. (2007), 'A united call to fight warming; Blair, Schwarzenegger push global leaders for action', *Washington Post*, 27 June, A14.

Krosnick, J. A., A. L. Holbrook and P. S. Visser (2000), 'The impact of the fall 1997 debate about global warming on American public opinion', *Public Understandings of Science* 9, 239–60.

McCright, A. M. and R. E. Dunlap (2000), 'Challenging global warming as a social problem: an analysis of the conservative movement's counter-claims', *Social Problems* 47, 499–522.

McCright, A. M. and R. E. Dunlap (2003), 'Defeating Kyoto: the conservative movement's impact on US climate change policy', *Social Problems* 50, 348–73.

Moore, M. (2007), 'Renewable energy: Dems, GOP diverge', *USA Today, 27 September*, www.usatoday.com/news/politics/election2008/2007-09-27-candidates-climate_N.htm [31 October 2007].

National Academy of Sciences (1991), *Policy Implications of Greenhouse Warming*, Washington, DC: National Academy Press.

Nelson, T. N., R. A. Clawson and Z. M. Oxley (1997), 'Media framing of a civil liberties conflict and its effect on tolerance', *American Political Science Review* 91, 567–83.

Orestes, N. (2004), 'Beyond the ivory tower: the scientific consensus on climate change', *Science* 306, 1686.

Parker, L. B., and J. E. Blodgett (1999), *Global Climate Change Policy: From 'No Regrets' to S. Res. 98*, www.ncseonline.org/nle/crsreports/climate/lim-17.cfm/ [1 November 2007].

Parker, L., and J. E. Blodgett (2007), *Global Climate Change: Three Policy Perspectives*, www.ncseonline.org/nle/crsreports/climate/clim-1.cfm#Introduction/ [1 November 2007].

PIPA (Program on International Policy Attitudes) (2005), *Overwhelming Majority of Americans Favors US Joining with G8 Members to Limit Greenhouse Gas Emissions*, www.pipa.org/OnlineReports/ClimateChange/ClimateChange05_Jul05/Climate Change05_Jul05_pr.pdf [2 November 2007].

Polling Report (2007a), *Environment*, www.pollingreport.com/enviro.htm [2 November 2007].

Polling Report (2007b), *Energy*, www.pollingreport.com/energy.htm [13 November 2007].

PRC (Pew Research Center for the People and the Press) (2007), *Global Warming: A Divide on Causes and Solutions*, pewresearch.org/pubs/282/global-warming-a-divide-on-causes-and-solutions [31 October 2007].

Ramseur, J. L. and B. D. Yacobucci (2007), *Climate Change Legislation in the 110th Congress*, www.ncseonline.org/NLE/CRS/abstract.cfm?NLEid=1883 [1 November 2007].

Roosevelt, M. (2007a), 'Regional pact caps emissions; leaders of six states and two Canadian provinces agree to cut output of greenhouse gases to 15 per cent below 2005 levels', *Los Angeles Times*, 23 August, B1.

Roosevelt, M. (2007b), 'Activists say gov. is green, but cautious; Schwarzenegger signed 19 key environmental bills this year. But his vetoes leave some observers disappointed with his record', *Los Angeles Times*, 22 October, B1.

Rosenbaum, D. (1993), 'Clinton backs off plan for new tax on heat in fuels', *New York Times*, 9 June, A1.

Sample, I. (2007), 'Scientists offered cash to dispute climate study', *Guardian*, 2 February, www.guardian.co.uk/ environment/ 2007/ feb/ 02/ frontpagenews. climatechange [2 November 2007].

Senate Report *S. Res. 98*, 105–54 (1997), frwebgate.access.gpo.gov/cgi-bin/ getdoc.cgi?dbname=105_cong_reports&docid=f:sr054.105.pdf [1 November 2007].

White House (2007), *President Bush Participates in Major Economies Meeting on Energy Security and Climate Change*, www.whitehouse.gov/news/releases/2007/ 09/20070928-2.html [1 November 2007].

Wigley, T. M., V. Ramaswamy, J. R. Christy, J. R. Lanzante, C. A. Mears, B. D. Santer and C. K. Folland (2006), *Temperature Trends in the Lower Atmosphere: Understanding and Reconciling Differences*, www.climatescience.gov/Library/sap/sap1-1/finalreport/sap1-1-final-execsum.pdf [2 November 2007].

Williams, C. J. (2007), 'Gov. says climate not political issue; Schwarzenegger lauds the Florida governor, and he bashes Bush and Detroit for what he calls inaction', *Los Angeles Times*, 14 July, A12.

Yacobucci, B. D. and L. Parker (2006), *Climate Change: Federal Laws and Policies Related to Greenhouse Gas Reductions*, www.ncseonline.org/NLE/CRS/abstract.cfm?NLEid=312 [1 November 2007].

Zaller, J. R. (1992), *The Nature and Origins of Mass Opinion*, New York: Cambridge University Press.

6
Hot Air and Cold Feet: The UK Response to Climate Change

Irene Lorenzoni, Tim O'Riordan and Nick Pidgeon

Setting the scene

Prior to the election of New Labour in 1997, the previous Conservative administration began a series of policy initiatives that inadvertently contributed to major reductions in UK greenhouse gas emissions. The then Prime Minister Margaret Thatcher was persuaded in 1988 that climate change posed dangers to both national security and the economy, and delivered landmark speeches to the Royal Society and at the United Nations. She also encouraged the formation of the Hadley Centre in 1990 and a decade later the Tyndall Centre, two of the world's premier climate research institutes. During the mid-1980s Thatcher also presided over a bitter dispute with the coal miners which culminated in the phasing out of much of Britain's coal production, followed in 1990 by the Conservative privatization of the UK electricity industry. These essentially political events precipitated a major shift in UK energy production from coal to low-price gas, allowing subsequent governments to claim that the UK was on course to meet its Kyoto target of reducing greenhouse gas emissions to 12.5 per cent below 1990 levels by 2008–2012. The demise of heavy engineering sectors in the early 1990s added to the UK's beneficial but serendipitous emissions reductions, while the introduction of an 'escalator' on road-fuel taxes in 1994 (later dropped) further added to the Tory contribution.

One of the hallmarks of New Labour's agenda was to engender a progressive environmental politics that included ecological care alongside social and economic considerations. Espousing an ecological modernization ideology (Weale 1992), the then Prime Minister Tony Blair declared that becoming greener was not antithetical to economic growth, but was both a business and a political opportunity, especially on the international scene (Jordan 2001). This was exemplified by Blair's steadfast promotion of the UK's international leadership on climate change and the injection into successive Labour Party manifestos of the aspiration that the nation would cut CO_2 emissions by 20 per cent against the 1990 baseline by 2010, well above the Kyoto target.

More than a decade later, the UK now faces a very different situation. Although the government has committed the UK to a 60 per cent cut in CO_2 emissions by 2050, in practical terms its energy and carbon budgets are in disarray. The 20 per cent cut in CO_2 emissions promised by 2010 looks increasingly untenable given that national emissions are rising by around one per cent annually (Cambridge Econometrics 2008). In part thanks to the 'fortuitous' events noted above (RCEP 2000: 83), the UK may just deliver its Kyoto target, although a recent report by the National Audit Office (NAO 2008) concludes that the national emissions accounts may be in error by over 15 per cent. Controversy also abounds on how appropriate levels of mitigation will be achieved. The muddle is further demonstrated by the government's championing of airport expansion and motorway widening, with only vague or delayed promises on vehicle pricing and carbon-neutral housing.

In this chapter we argue that the UK is suffering from a severe mismatch between good political intentions and reality. A number of factors account for this; however, we focus in particular on structural inadequacies within the UK's political system of environmental governance, coupled with a failure of the ecological modernization thesis' emphasis on purely technological and market-based solutions. We do not provide a detailed critique of the main climate policy instruments used in the UK, as these have already been extensively discussed in the literature (see, for example, The Royal Society 2002; Sorrell 2003; Ekins and Etheridge 2006). However the evidence presented in Table 6.1, which provides a summary of the instruments deployed, supports our overall argument that UK climate policy has been structurally constrained by the government's allegiance to ecological modernization and reliance on market mechanisms that focus on producing incremental changes in business emissions.

Developments in UK climate policy

Despite the Conservative government's attention to climate change in the mid-1990s, it was the Labour government that catapulted the UK's contribution to addressing it onto the international scene and introduced the majority of the policy instruments that make up the mainstay

Table 6.1 Key UK and EU climate policy instruments

Date	Policy instrument	Description	Commentary
Energy			
1990	Non-Fossil Fuel Obligation (NFFO)	Legal obligation on electricity generators to supply a fixed percentage of electricity from non-fossil fuels; funded investment in non-fossil fuels	Indiscriminate in terms of non-fossil fuels supported: majority of early funding to nuclear power; short time horizons undermined investor confidence; near-market technologies favoured over 'blue skies' innovation
2002	Renewables Obligation (RO)	Obliges electricity suppliers to source a growing proportion of electricity supplied from renewable sources as certified by Renewables Obligation Certificates (ROCs)	ROCs' price provides inadequate and costly incentive for long-term investments in renewables; does not address 'bottlenecks' created by planning system (Connor 2003)
2002	Energy Efficiency Commitment/ Carbon Emissions Reduction Target (CERT)	Mandatory targets for gas and electricity suppliers with 15,000 or more customers to assist customers to improve domestic energy efficiency	Helps to address fuel poverty by requiring suppliers to achieve at least half of energy savings in households on income-related benefits and tax credits
Industry/commerce			
2001	Climate Change Levy (CCL)	Tax on business energy use, offset by cuts in employers' National Insurance Contributions (NICs), and support for investment in energy efficiency and renewables	Criticized for not taxing carbon directly and for creating negligible financial incentive for long-term investment; NIC rebates favour service industries over energy-intensive sectors (The Royal Society 2002)

2001	Climate Change Agreements (CCAs)	80% discount in CCL for energy-intensive and (from 2004) trade-exposed sectors which meet binding emissions/energy efficiency targets	Target setting too lax and not made more stringent following 2004 review; however created significant initial awareness-raising effect (Ekins and Etheridge 2006)
2002	UK emissions trading scheme (UK ETS)	Voluntary emissions trading by businesses, with incentive fund for non-CCA participants	Created loophole for companies to avoid loss of reduced-rate CCL; overlap with CCL, CCAs and EU ETS; limited trading and lessons for EU emissions trading (Sorrell 2003)
2005	EU emissions trading scheme (EU ETS)	Mandatory emissions trading for sectors covered by EU emissions trading directive	Generous allocations in most member states in Phase 1; tightened in phases 2 and 3 (see EU chapter)
2007	Carbon Reduction Commitment (CRC)	Mandatory emissions trading for large non-energy intensive commercial and public organizations	Consultation in 2006 showed strong support for CRC to be mandatory rather than voluntary. To be introduced under enabling powers planned in the Climate Change Bill

of the UK's emissions reduction climate strategy. These instruments and the general strategy underpinning them were driven in large part by the Royal Commission on Environmental Pollution (RCEP) report of 2000. Building on the premise that anthropogenic climate change could be detrimental for Britain and the rest of the world, the RCEP advocated that the UK should cut its CO_2 emissions by 60 per cent by 2050, arguing that (i) developed nations have a moral duty to reduce emissions, with a view to encouraging developing countries also to engage in climate change mitigation; (ii) such targets could support the international climate regime; and (iii) the UK should bear the brunt of reducing emissions now to avoid the greater costs of future impacts. The key issue identified by the RCEP was that should demand not be curtailed, mitigation will depend upon carbon neutral energy generation.

Prior to the publication of the RCEP findings, much government attention had been given to the impacts of climate change. Work in this area has been coordinated by the UK Climate Impacts Programme (UKCIP), which, in collaboration with the Met Office and the Hadley Centre, produced landmark national climate-change scenarios in 1998 and 2002. Updated ones based more directly on risk assessments were scheduled for late 2008. The RCEP's report also coincided with extensive flooding throughout the UK, which served to increase public awareness of the climate issue (Jordan 2001). These developments opened the floodgates to a plethora of reviews, policy documents and initiatives aimed at building the UK's climate change mitigation portfolio.

In its Climate Change Programme of 2000 (DETR 2000), the government stated its commitment to a low-carbon economy and the need to review UK energy generation and consumption, including the feasibility and acceptability of different energy provision options. The government suggested that nuclear power might become a major contributor to climate change mitigation, although it conceded that cost considerations, uncertainty about waste disposal and public reservations over nuclear safety suggested that further debate was required before new nuclear power stations could gain favourable and 'fast track' planning treatment.

By 2003, however, the government acknowledged that the UK's energy policy needed major restructuring in response to ageing energy infrastructure, declining North Sea oil and gas supplies, changes to the supply mix and energy provision, and climate change. The 2003 Energy White paper enshrined the government's main priorities: a commitment to significant mitigation (enshrining in policy the RCEP's 60 per cent CO_2 reduction target, with the interim goal of curbing CO_2 emissions by 20 per cent below 1990 levels by 2010), and maintaining reliability of energy supplies through a combination of renewable energies, energy efficiency and reduced energy consumption. Although it stated that the nation's future energy mix would be decided through market mechanisms, it also made clear that nuclear could be an attractive option, despite at that time being economically non-viable, if it was deemed 'necessary' to meet carbon targets (DTI 2003: 12). Although the White Paper stressed that any decision would be preceded by full public consultation and another White Paper setting out the government's proposals, it nevertheless signalled the political reframing of nuclear power as an important part of the UK's potential climate change mitigation strategy (Pidgeon *et al.* 2008).

2006 heralded a further shift in the underpinnings of UK energy policy. Despite pre-election promises, the new UK Climate Change Programme announced that the nation would narrowly fail to achieve its interim target of a 20 per cent cut in CO_2 emissions by 2010 and instead would achieve a cut of only 15–18 per cent. Among the range of measures announced was a new Energy Review, barely three years after the previous one (HM Government 2006). That document also announced that the economic viability of nuclear power had been improved by rising fuel prices and the carbon-pricing effects of the EU Emissions Trading Scheme (EU ETS), and that use of nuclear power had the dual purpose of delivering on climate change mitigation and energy security.

The revised Energy Review was severely criticized by some government committees and NGOs, with the media and various stakeholders arguing that the Prime Minister's personal support for nuclear power had made the outcome of the consultation a foregone conclusion. Greenpeace and several other NGOs even applied for a judicial review on the grounds that information on waste disposal and the economics of nuclear power had not been available (the final report of the Committee on Radioactive Waste Management recommending the long-term option of geological storage for radioactive waste was published only after the consultation had concluded in June 2006), and that the Review's focus mainly covered electricity provision, which accounts for only one-third of UK emissions.

In a landmark decision in February 2007 the High Court ruled procedural unfairness and breach of expectation. The government accepted the verdict, which meant that it was required to repeat the consultation. As a result, the *Energy White Paper* (DTI 2007) of May 2007 outlined plans to reduce reliance on fossil fuels by increasing sustainable energy output by 2015, reducing energy use, and placing greater emphasis on local energy production, renewables and new technologies, while also launching a revised consultation on new nuclear build. In so doing, the government reiterated the potential contribution of nuclear power to meeting national emissions targets and its interest in reaching a decision promptly, given previous delays and the long lead times should new nuclear build be approved (Hines 2007).

Despite a critical assessment by the government's own independent sustainability advisers, the UK Sustainable Development Commission (SDC 2006), as well as further external critique, in early 2008 the government appeared to map out the policy foundations for a new generation of nuclear power stations (BERR 2008). However whether these stations will be constructed is uncertain, unless very favourable (and effectively subsidized) financial and planning arrangements are established and a disposal solution is found that can accommodate both the legacy wastes from earlier operations and the new wastes from a future generation of nuclear power stations.

Alongside these deliberations, other events – both national and international – have exerted major influences on UK climate policy (see Table 6.2). One obvious example is the EU ETS (see Damro and MacKenzie, this volume, for a detailed discussion of this scheme). At

Table 6.2 Timeline of key UK and international climate change policies and events

1988	PM Thatcher speech to Royal Society
1989	Thatcher's speech on global environmental change to the United Nations
1990	First IPCC Report; Hadley Centre for Climate Prediction and Research opens
1992	UN Framework Convention on Climate Change signed
1995	IPCC Second Assessment Report; COP-I Conference in Paris
1997	Labour promises to cut national CO_2 emissions by 20% by 2010 compared with 1990 levels
	Kyoto Protocol agreed. UK's target: reduction of emissions by 12.5% by 2008–12 relative to 1990
1998	Energy review (DTI): temporary moratorium on gas-fired power stations
2000	RCEP advocates 60% cut in CO_2 emissions by 2050 relative to 2000 levels; first UK Climate Change Programme; UK Climate Impacts Programme (UKCIP) set up with focus on adaptation; decline in North Sea gas and oil begins with UK poised to become net importer; launch of Tyndall Centre for Climate Change Research
2001	IPCC Third Assessment Report; Climate Change Levy (CCL) and Climate Change Agreements (CCAs) introduced

2002	Energy Review (PIU): 'radical agenda' to create powerful incentives to decarbonize domestic energy supplies, pursue energy needs and achieve 'environmentally sustainable energy system'; UK Emissions Trading Scheme launched; Building Regulations updated; first phase of the Energy Efficiency Commitment (EEC) (2002–2005); Renewables Obligation (RO) launched
2003	Energy White paper aimed at energy security, climate change, fuel poverty; Aviation White Paper
2004	Climate Change Communications Strategy commissioned by DEFRA; Energy Efficiency Action Plan; Future of Transport White Paper
2005	EU ETS commences; second phase of EEC (2005–2008); G8 Gleneagles Summit puts climate change on the agenda
2006	Revised Climate Change Programme; Energy Review (DTI) re-emphasizes economic case for new nuclear build; Stern Review (Chancellor and Treasury) argues that it makes economic sense to mitigate now; establishment of Climate Challenge Fund; new Building Regulations set standards for new buildings
2007	EU integrated energy and carbon policy (unilateral 20% CO_2 emissions cut by 2020 from 1990); IPCC Fourth Assessment Report warns that 'warming of the climate system is unequivocal'; draft Climate Change and Energy Bills; Energy White Paper (DTI) announces Carbon Reduction Commitment Scheme; second consultation on government's arguments for new nuclear build; EU Green and White Papers on adaptation; voluntary agreement with producers to cease production of incandescent lightbulbs by 2011; consultation on reform of Renewables Obligation; Planning White Paper
2008	Nuclear Power White Paper; negotiations to include aviation in EU ETS; consultation on Carbon Reduction Commitment Scheme and Climate Change Simplification project to simplify relationship between EU ETS, CCAs and CRC; third phase of EEC (2008–2011) renamed Carbon Emissions Reduction Target (CERT); RCEP review of adaptation in the UK; UKCIP08 climate scenarios published

the end of 2006 the *Stern Review of the Economics of Climate Change* (Stern 2006) became an important international benchmark and was endorsed nationally by the Chancellor and Prime Minister. The Stern report was significant not only because of its precautionary message but also because it was commissioned by the Treasury, the most powerful Ministry within Whitehall. Extolling the economic benefits of early mitigation, Stern effectively reiterated the message of the RCEP (2000) and Intergovernmental Panel on Climate Change (IPCC 2007) in language that captured the attention of aware and concerned public and business sectors worldwide (Jordan and Lorenzoni 2007). In June 2007 the G8 Summit was also instrumental in putting environmental issues at the centre stage of politics and public debate (Hale 2007), while at grass-roots level the government's Climate Change Communication Initiative and associated fund were established to encourage regional responses to climate change.

In late 2007 the new Prime Minister Gordon Brown announced proposals to include economic estimates of impacts that could arise from climate change in policy and investment decisions (DEFRA 2007), thereby enacting the most salient recommendations of the Stern Review. However there are doubts about the influence of the current carbon price on the viability of energy and transport schemes (Ekins 2008). The UK also led debates to include the aviation sector in the EU ETS, suggesting that action on the international scene is taking precedence over national priorities, especially given that progress on mitigation has been slowest in the transport sector, in part due to inter-departmental conflicts (Carter and Ockwell 2007; Jordan and Lorenzoni 2007).

In the wake of the Stern report, the government announced its Climate Change Bill. If passed by Parliament it would mandate the 60 per cent reduction in CO_2 emissions by 2050 and create an intermediate target of a 26–32 per cent reduction by 2020 (excluding international aviation and shipping). The Bill also proposes five-year carbon budgets, with annual progress reporting by Ministers. Government would be assisted by advice from an independent Committee on Climate Change and, importantly, the Bill would enable ministers to introduce additional measures without need for legislation.

Several core provisions brought forward in the Climate Change Bill had their origins in earlier developments outside of Cabinet Government. Over 400 of the 646 Members of Parliament signed an early day motion after the 2005 general election petitioning for a Climate Change Bill (based upon a Private Members Bill drafted earlier that year by Friends of the Earth). A joint statement was subsequently agreed in early 2006 by the main opposition parties calling for a political consensus on targets. This was followed by a report commissioned by the All Party Parliamentary Climate Change Group outlining the capacity of cross-party consensus to free long-term planning from the pressures of the electoral cycle (Clayton *et al.* 2006). It also observed that cross-party agreement on *targets* would still allow room for legitimate political disagreement around the *means* for meeting targets.

Others have argued, however, that the Bill disregards recent developments in climate science, and that emissions should be cut by at least 80–90 per cent by 2050 (Anderson and Bows 2007). The Bill also failed to commit the government to annual emissions-reduction targets (many

NGOs advocate three per cent per annum) on the grounds that these would be susceptible to the effects of short-term fluctuations in energy prices and weather. NGOs contested this, arguing that the five-year budgets – being longer than the *de facto* four-year electoral cycle – could undermine the electoral accountability of government on this issue.

These comments notwithstanding, the Climate Change Bill is an original and forward-looking proposal because, if it is approved, the UK would become the first country to make long-ranging and ambitious mitigation targets legally binding, and because its existence is the result of extensive cross-party agreement on climate change. Although the momentum generated fuelled expectations, however, and Gordon Brown's budget in March 2007 confirmed some forward-looking policies, such as stricter building standards to ensure all new housing is carbon neutral, many observers were disappointed. Although the UK can be praised for its influence and leadership on climate change internationally, and for moves towards apparent party consensus, more attention needs to be dedicated towards actually enacting effective and significant actions at the national level (Clayton *et al.* 2006).

Obstacles to strengthening national climate action

To date the UK government's aspirations of initiating a transition to a low-carbon economy have focused predominantly on (i) restructuring and diversification of future energy supply options, embedded within considerations about energy security and reliability but increasingly reframed in terms of climate change mitigation (Pidgeon *et al.* 2008); (ii) market mechanisms for reducing emissions from business sectors (see Table 6.1); and (iii) information diffusion to encourage attitudinal and behavioural change by means such as encouraging or mandating new standards, more energy-efficient products and better labelling, and by direct 'social marketing'.

Three main features characterize these policies. First, they are underpinned by the government's steadfast adherence to the paradigm of ecological modernization, something that we would argue strongly undermines the political will for radical thinking and action. Second, even if such political will did exist, action at a national level is hampered by ingrained structural inadequacies in the UK system of Cabinet Government, and by a degree of unevenness in decision-making and planning policy brought about by the systems of devolved administration in Scotland, Wales and Northern Ireland. Third, there is a failure to reflect emerging evidence of public concern about climate change and calls for strong political leadership in national policy-making. Indeed the UK could be described as facing a climate governance 'trap' in that public(s) are increasingly expressing a desire for strong action whilst politicians remain wary of bold steps for fear of short-term electoral retribution.

Notions of ecological modernization – the theory that environmental protection can develop hand-in-hand with economic growth through the development and use of appropriate technologies – were clearly outlined by Tony Blair in 1997, and in 2008 remain prevalent amongst most government officials and business stakeholders. The Stern report sounded a strong caution to this, arguing that market failures would be likely to undermine traditional economic policies towards innovation, in particular investments in low-carbon technologies. Despite this and numerous government documents acknowledging the importance of addressing climate change cohesively and through multiple strategies, technological solutions and voluntary agreements have remained dominant (Carter and Ockwell 2007). Indeed Labour has been accused of prioritizing economic considerations and business interests over environmental ones. For example, ministerial approval in 2008 of a new coal power station at Kingsnorth in Kent was justified on the grounds that it is needed to meet projected energy shortfalls despite the fact that no commitment had been made to equip this station for carbon capture and storage (Macalister 2008). There is also a persistent and influential business lobby which slows innovation and radical interference in markets and investment. For example, the mandating of carbon-neutral homes, which is easily achievable at reasonable cost, is being delayed until 2016 because of pressure from the construction lobby and appliance manufacturers.

The failure to get to grips with carbon emissions is also due in part to fears that to do so would lead to a significant loss of international competitiveness, especially in relation to China. There is also an important European dimension to this argument, given the lack of coherent EU strategy to meet its target of reducing emissions by 20 per cent by 2020, as this has nourished a discourse according to which the UK cannot go it entirely alone without compromising its export base. Although there is a clear case for new innovation in the clean technology sector, Cambridge Econometrics (2008) concluded that the UK will only reach a five per cent contribution of renewables to its energy mix by 2020, rather than the legally binding 15 per cent required by the EU. This is because of excessively perverse pricing and regulatory arrangements which penalize this sector at the expense of the heavily subsidized nuclear sector, and even arguably promising new technologies relating to carbon storage and sequestration are not being sufficiently encouraged by fiscal incentives. By contrast, the German, Danish and Swedish renewables sectors are flourishing and gaining real profits from feeding spare power into their grids.

Structural factors rooted within the UK's system of Cabinet Government also obstruct vigorous action on climate change. There is no coherent system of policy coordination in the Cabinet structure that prioritizes climate change as a strategic economic, social, environmental and institutional issue. The Economy and the Environment Cabinet Committee remains in the Treasury and strongly influenced by the business enterprise culture favoured by ecological modernists. The core environmental aspects of climate policy, meanwhile, reside in the Department of Environment, Food and Rural Affairs (DEFRA), a weak ministry that has been buffeted in recent decades by crises such as BSE, controversies over GM crops, and foot-and-mouth disease. Energy and innovation policy – two key routes to mitigation – by contrast are controlled by the former Department of Trade and Industry (now Department for Business, Enterprise and Regulatory Reform), a department with close links to the Treasury and a singular history of lack of cooperation with DEFRA. Finally, transport policy and the domestic sector fall within other Ministerial portfolios. Although the new Prime Minister recently outlined his commitment to a long-term approach to mitigating climate change (Brown 2007), the first months of his mandate do not indicate that the environment is as high a priority as it was for Blair (Carter and Ockwell 2007).

These departmental disjunctions have perpetuated disjointed governance of a cross-cutting issue *par excellence*. Strong lobbying by corporate and other interests have, as a result, led to watered-down or inconsistent policies. There is considerable tension, for example, between the Climate Change Bill objectives and energy and transport policies. Other considerations, including energy reliability and security, have also strongly influenced the interface between climate and energy policy. The differing geographical scales of analysis and action required for climate change mitigation and adaptation can also complicate the message. Grassroots disillusionment with the government's failure to deliver on climate change and sustainable development has spurred a plethora of local initiatives (for example, Transition Towns and Carbon Reduction Action Groups) aimed at taking matters into their own hands by developing carbon neutral sustainable communities.

Inconsistencies among climate policies are exacerbated by the ongoing political processes of devolution in Scotland, Wales and Northern Ireland. Tensions tend to occur when powers and aspects of policy which are fully or in-part devolved, such as aspects of environmental protection or planning, conflict with others which are reserved for national government, such as national energy strategy. For instance, in spite of the enthusiasm in government circles south of the border, the current Scottish administration will not pass legislation authorizing any new nuclear construction.

In other respects devolution is bringing opportunities for change at regional level. Scottish politics is traditionally more committed to social justice considerations, and as a result bus transport is free for pensioners and subsidized for families on social benefits. In addition, social housing is subsidized for energy efficiency measures in Wales and Scotland to a far greater extent than in England. And in both Scotland and Wales planning guidelines are more favourable to the siting of renewables, especially wind power, compared with England, where over 80 per cent of applications are opposed. In institutional terms, and unlike the UK government, the Welsh Assembly Government aspires to an annual carbon reduction target of three per cent per annum by 2011, and has established a Climate Change Commission for Wales: a body with broad cross-party, business and NGO membership and support and a remit to develop consensus policies. In essence, the Wales and Scottish legislatures and their business, environment and audit/regulatory counterparts appear to be more cohesive in relation to climate policy than is the case in Westminster, and look set over the coming years to exploit their respective geographical advantages in wind and tidal power (SDC 2007).

The third obstacle to progress concerns the failure to link evidence of public support for action and guidance on climate change to bold political strategies. Whilst Tony Blair provided rhetorical leadership at the international level, his impact on the domestic scene was limited, a pattern that appears to be repeated with his successor, Gordon Brown. Although government acknowledges the substantial individual contributions people make to climate change through lifestyle and daily practices, it has shied away from substantial interventions to change these. In particular, it has failed to take advantage of opportunities for reducing emissions from the domestic sector despite there being a range of easily achievable measures (Carter and Ockwell 2007). The government's communication strategies and its emphasis on voluntary uptake have yet to lead to substantial emissions cuts, let alone long-term behavioural or attitudinal change (Lorenzoni *et al.* 2007). The Carbon

Trust is trialling a carbon label designed to encourage business and consumers to reduce their carbon footprints, while DEFRA (2008) has identified twelve key everyday activities and behaviours that should be the target of communications and other interventions, including installation of domestic insulation, better home energy management, less waste of food, use of more efficient vehicles and fewer unnecessary flights. However research in other domains (Jackson 2005; Maio *et al.* 2008) has shown that information provision alone does not necessarily result in changed attitudes or behaviours. Instead such changes require bold interventions aimed at, amongst other things, breaking environmentally damaging habits, providing fiscal and other incentives, addressing the current 'moral climate' or social norms that for most still privilege consumption over environmental protection, and making major changes to infrastructure that help people to adopt and sustain low-carbon lifestyles.

There is growing and increasingly consistent evidence on how the public in the UK and other developed countries views climate change. Public views reflect, at least in part, media and policy discourses which, in relation to climate change, are constantly evolving. However whereas there is an apparent consensus in the UK on the existence of anthropogenic climate change, Ereaut and Segnit (2007: 6) maintain that 'weak spots' in the discourses underlying this consensus make it vulnerable and hence argue that it is therefore vital to capitalize on the present momentum by developing communications that truly engage citizens' behaviour with climate change. Numerous polls and psychological and perception studies report almost universal awareness and concern about climate change among the UK public, yet few individuals have taken substantial measures to reduce their impact despite a declared willingness to do so. A number of characteristics make climate change particularly difficult to conceptualize. Most people perceive climate change as spatially and temporally removed from their daily priorities (Lorenzoni and Pidgeon 2006; Lorenzoni *et al.* 2007). Although most can identify the large-scale impacts expected, fewer have accurate knowledge of the causes of, and solutions to, climate change (Lorenzoni *et al.* 2006). As a key source of information, the value of media translation of scientific findings through confusing and contradictory messages, alarmist (and sometimes optimistic) reporting, use of loaded terminology and dramatic imagery has been called into question (Ereaut and Segnit 2006). Rather than inspiring action, exaggerating the dangers without indicating how they might be avoided can induce feelings of guilt and hopelessness (Moser and Dilling 2004; Jowit 2007).

Furthermore, the evidence suggests that individuals in the UK population tend either to ascribe responsibility for addressing climate change to others (the majority), to national governments or to the international community, or feel responsible but constrained from being able to reduce their personal emissions by the societal systems within which they operate. In the latter case, removing structural barriers may enable some individuals to act. The former, however, suggests the need for consistent and significant long-term policy responses which reinforce an acknowledgement of personal responsibility for individual actions and duty towards others and the environment, and hence provide support for acting on this basis. Ensuring a sustainable future could even be considered, in the future, to be an inherent individual moral duty.

Personal carbon allowances (PCAs) have been proposed as one political strategy that could foster this and counteract the relentless rise of domestic and transport emissions. They attracted the interest of David Miliband, then Secretary of State for the Environment, and in 2008 the House of Commons Environmental Audit Committee (2008) urged the government to intensify investigations into how PCAs could be made publicly and politically acceptable. Under a PCA scheme, individuals would be allocated carbon credits which they would use for necessary energy purchases (heating, electricity). Those who use less than their allocation could then sell their surplus credits to higher carbon users. Although doubts have been raised about the practicalities of a PCA scheme on the grounds of complexity, controllability, monitoring, leadership and accessibility, it has been portrayed as a fair and equitable means of engaging individuals and communities in daily behavioural change which carbon taxation, for instance, would not engender. Although it is argued that public opposition to PCAs is likely to be strong, given their direct targeting of lifestyles, the mechanism may become more acceptable if indeed post-2012 negotiations do include some form of mitigation targets for major developing countries (Seyfang *et al.* 2007).

Conclusions and ways forward

In this chapter we have sought to demonstrate that despite strong climate policy rhetoric, political leadership on the international stage and world-class climate research centres, the UK's record on the implementation of national climate policies has been less than impressive. In this section we summarize the reasons for this and discuss options for reducing the current mismatch between the government's climate policy rhetoric and achievements.

The first problem is the government's predisposition towards ecological modernization and market mechanisms that focus on incremental change, and its tendency to shy away from bolder actions that politicians may believe (in some cases incorrectly) will be resisted by corporate sectors or the electorate. Investment in technology and infrastructure is also hampered, *even* assuming that ecological modernization can be made to work, by a lack of mechanisms that reflect the true external costs of carbon and by the Treasury's adherence to discount rates which penalize long-term investment even where these investments are in the national economic interest (SDC 2007). Strong targets have also been consistently undermined by ineffective policy measures, poor reporting, auditing reviews which are persistently ignored, and repeated upward revisions in carbon emissions scenarios.

Second, progress has been hampered by departmental interests and interdepartmental conflicts that have been compounded by a lack of coherent policy coordination in a UK Cabinet structure that does not make climate change a strategic economic, social, environmental and institutional priority, and by the added decoupling of national intentions from practical actions caused by devolution.

A third factor has been the disjunction between the government's weak positioning on climate change and emerging public concern about climate change and commitment to reducing its own carbon emissions. Government concerns about energy insecurity have led it to do almost anything to promote new energy supply, even coal, overlooking a growing wish within senior sections of UK businesses, and among the general public, for more steadfast policies to promote energy efficiency and carbon saving. One reason why these are not being implemented is an ideological unease about interfering with people's behaviour. Overall, however, the most significant reason for the non-implementation of climate politics in the UK lies in the weakness, even in the face of a majority Parliament and growing business and civil society concerns, of modern 'democracy' to force government to be bold.

The previous sections also outlined how UK mitigation policies are subject to growing concerns not only about climate impacts but also about ensuring energy self-sufficiency and security, economic growth and mobility. In this respect one might even speculate that the discourse of climate mitigation is itself being reframed within government to promote economic and political priorities. In addition, the government has adopted a diversity of indirect mechanisms to promote climate considerations. The Climate Change Bill, though a welcome step, is an example of risk distribution through cross-party agreement and reliance on an independent committee for decisions that could cause resistance or friction. Equally, over-reliance on 'communication' as a stimulus for behaviour change sidesteps the awkward issue of mandating direct interventions to reduce personal carbon use. Examples of successful government interventions abound in other fields, for example legislation mandating vehicle seat restraints, and the ban on smoking in public houses in Britain. In both cases resistance was overcome not by changing attitudes first but by upstream interventions to force changes in behaviour in the (correct) anticipation that this would in turn lead to modified downstream attitudes. Our contention is that climate policy urgently needs to throw off its timidity and take heed of such lessons.

Beyond government spheres the climate debate is widening to include attention to adaptation (Pielke *et al.* 2007). The Stern Review put adaptation back onto the agenda, and the RCEP is poised for a major review of adaptation in the UK. Increasingly adaptation and mitigation policies will need to be linked. The summer flooding in 2007 underscored the UK's vulnerability to extreme weather events, and public interest is now focused on the need to be better prepared. Again, this may create a window of opportunity to strengthen climate policies by integrating adaptation and mitigation considerations, as has been advocated by the SDC.

Successful climate change policies are also likely to be those that involve detailed and critical examination of the UK's social, technical, institutional and infrastructure characteristics, with a view to creating a more equitable, just, healthy and low-impact society. Significantly reducing the impacts of climate change implies long-term commitments to reform economic and social policy. The government's approach to climate policy must also focus both on adaptation in the short term and mitigation in the long term. Above all a strong emphasis should be placed on government putting its own house in order so as to ensure a coherent and joined-up response across ministerial departments to the demands (which may ultimately prove incompatible) of a competitive global economy and national obligations to the international climate process. We are not suggesting that the UK ignore international signals – this would be unrealistic, and global interconnectedness is in no small part the cause of current emission problems. However, enacting strategies at the national and sub-national levels now deserves attention. As we have highlighted above, a politically feasible strategy would be to cement and extend wherever possible the existing cross-party consensus on climate change beyond targets to include the means to meet those targets. The very existence of the Climate Change Bill, as well as developments within the devolved administrations, suggests that this may be possible. The main instrument of the bill, the independent Climate Change Committee, might also be asked to work with business and regulators to set prices, taxes and incentives to ensure a transition to a low-carbon economy. And if the commitment to market mechanisms is to be retained, this requires a consistent set of signals for appropriate pricing trajectories, as well as regulatory incentives to charge properly for carbon reduction technology and management.

Bridging the gap between climate policy and implementation also requires a better understanding of the mechanisms and relative success of grassroots initiatives, coupled with creative means to remove barriers inhibiting behavioural change. One concrete possibility would be to begin 'live' community experiments with personal carbon allowance schemes so that practicalities and social justice issues can be explored before any large-scale implementation. Another is to encourage cohesive community-based initiatives, particularly on the issue of buildings and sustainability. To date the government's approach has been *ad hoc* – it is time the electorate is respectfully involved in initiatives which foster genuine ownership, pride and contribution to a better future. Part of the problem in engaging people with climate change is its long-term, complex and diffuse nature. Expressing concerns about climate change in ways that resonate with people's (moral) norms and values, even if this means transcending scientific uncertainty by couching it in terms of environmental, social and economic sustainability, is part of the challenge. The other is to enable people to act: facilitating attitudinal and behavioural change in tandem, and recognizing the complex linkages between them, can enable the transition to a low-carbon society. Technologies and approaches which genuinely embed within everyday activities, and which reveal linkages between behaviour and the environmental damage being caused, will also have an important role to play.

To be bold, there needs to be a new vision of what a low-carbon sustainable society and economy should look like and a clearer view of the changes in policy and institutional design necessary to get there. Moves to create such a vision and debate are currently not on the cards but, if they are not attempted, there will be no mechanism for bringing the public, private and civil sectors together for a common endeavour to address climate change. There is increasing evidence of a yearning in the country for such a bold vision and implementation plan. The science tells us that we have at best 15 years before irreversible change takes place, so we do not have the luxury of procrastinating. This is why influential commentators are beginning to look anew at the relationship between democracy and sustainability (see SustainAbility 2008). Above all, we have yet to devise a democracy which votes for uncertain, but possibly vital, future pathways to maintain a sustainable planet for all. Achieving such a democracy must involve a genuine dialogue with electors and civil society on terms that neither patronize them nor minimize the considerable stakes involved.

Acknowledgements

The preparation of this chapter was supported by a Leverhulme Trust award to Cardiff University and the University of East Anglia (No. F/00 407/AG). We wish to thank Hugh Compston for very helpful comments on an earlier version.

References

Anderson, K. and A. Bows (2007), *A Response to the Draft Climate Change Bill's Carbon Reduction Targets*, Tyndall Centre Briefing Note 17, University of Manchester.

BERR (Department for Business, Enterprise and Regulatory Reform) (2008), *Meeting the Energy Challenge: A White Paper on Nuclear Power*, London: BERR.

Brown, G. (2007), 'Speech by The Prime Minister The Right Honourable Gordon Brown MP on Climate Change hosted by the WWF at the Foreign Press Association, Monday 19 October 2007', Downing Street: Press Notice.

Cambridge Econometrics (2008), *UK Energy and the Environment*, Cambridge: Cambridge Econometrics.

Carter, T. and D. Ockwell (2007), *New Labour, New Environment? An Analysis of the Labour Government's Policy on Climate Change and Biodiversity Loss*, Report for Friends of the Earth, Centre for Ecology, Law and Policy, University of York.

Clayton, H., N. Pidgeon and M. Whitby (2006), *Is a Cross-Party Consensus on Climate Change Possible – or Desirable?* Report of First Inquiry to the All Party Parliamentary Climate Change Group, London: APPCCG.

Connor, P. (2003), 'UK renewable energy policy: a review', *Renewable and Sustainable Energy Review* 7, 65–82.

DEFRA (Department for Environment, Food and Rural Affairs) (2007), *Updated Guidance on the Shadow Price of Carbon*, www.defra.gov.uk/environment/climate change/research/carboncost/index.htm [7 January 2008].

DEFRA (2008), *A Framework for Pro-Environmental Behaviours*, London: DEFRA.

DETR (Department of Environment, Transport and the Regions) (2000), *Climate Change: The UK Programme*, London: DETR.

DTI (Department of Trade and Industry) (2003), *Energy White Paper: Our Energy Future – Creating a Low Carbon Economy* , London: DTI.

DTI (2007), *Energy White Paper: Meeting the Energy Challenge*, London: DTI.
Ekins, P. (2008), *Path of Least Resistance*, www.guardian.co.uk/environment/2008/feb/13/carbonemissions.travelandtransport [10 March 2008].
Ekins, P. and B. Etheridge (2006), 'The environmental and economic impacts of the UK climate change agreements', *Energy Policy* 34, 2071–86.
Ereaut, G. and N. Segnit (2006), *Warm Words. How Are We Telling the Climate Story and Can We Tell it Better?* London: IPPR.
Ereaut, G. and N. Segnit (2007), *Warm Words II. How the Climate Story is Evolving*, London: Energy Saving Trust and IPPR.
Hale, S. (2007), *A Greener Shade of Blue? Reflections on New Conservative Approaches to the Environment*, London: Green Alliance, pp. 2–8.
Hines, N. (2007), 'Darling dismisses "daft" anti-nuclear lobby and unveils energy strategy', www.timesonline.co.uk/tol/news/uk/article1828492.ece [20 July 2007].
HM Government (2006), *Climate Change. The UK Programme 2006*, Norwich: HMSO.
House of Commons Environmental Audit Committee (2008), *Personal Carbon Trading*, Fifth Report of Session 2007–08, London: The Stationery Office.
IPCC (Intergovernmental Panel on Climate Change) (2007), *Climate Change 2007: The Physical Science Basis. Summary for Policymakers*, Geneva: IPCC.
Jackson, T. (2005), *Motivating Sustainable Consumption: A Review of Evidence on Consumer Behaviour and Behaviour Change*, University of Surrey: Centre for Environmental Strategy.
Jordan, A. (2001), 'A climate for policy change? The contested politics of a low carbon economy', *The Political Quarterly* 72, 249–54.
Jordan, A. and I. Lorenzoni (2007), 'Is there now a political climate for policy change? Policy and politics after the Stern Review', *The Political Quarterly* 78, 310–19.
Jowit, J. (2007), 'Don't exaggerate climate dangers, scientists warn', www.guardian.co.uk/environment/2007/mar/18/theobserver.climatechange [4 January 2008].
Lorenzoni, I. and N. Pidgeon (2006), 'Public views on climate change: European and USA perspectives', *Climatic Change* 77, 73–95.
Lorenzoni, I., A. Leiserowitz, M. Doria, W. Poortinga and N. Pidgeon (2006), 'Cross national comparisons of image associations with "global warming" and "climate change" among laypeople in the United States of America and Great Britain', *Journal of Risk Research* 9, 265–81.
Lorenzoni, I., S. Nicholson-Cole and L. Whitmarsh (2007), 'Barriers perceived to engaging with climate change among the UK public and their policy implications', *Global Environmental Change* 17, 445–59.
Macalister, T. (2008), '£1bn coal-fired power station gets green light', www.guardian.co.uk/environment/2008/jan/04/fossilfuels.carbonemissions [4 January 2008].
Maio, G., B. Verplanken, A. Manstead, W. Stroebe, C. Abraham, P. Sheeran and M. Conner (2008), 'Social psychological factors in lifestyle change and their relevance to policy', *Social Issues and Policy Review* 1, 99–137.
Moser, S. and L. Dilling (2004), 'Making climate hot: communicating the urgency and challenge of global climate change', *Environment* 46, 32–46.
NAO (National Audit Office) (2008), *UK Greenhouse Gas Emissions: Measurement and Reporting*, London: NAO.

Pidgeon, N., I. Lorenzoni and W. Poortinga (2008), 'Climate change or nuclear power – no thanks! A quantitative study of public perceptions and risk framing in Britain', *Global Environmental Change* 18, 69–85.

Pielke, R. Jr., G. Prins, S. Rayner and D. Sarewitz (2007), 'Lifting the taboo on adaptation', *Nature* 445, 597–8.

RCEP (Royal Commission on Environmental Pollution) (2000), *Energy: The Changing Climate*, 22nd report, London: Stationery Office.

SDC (Sustainable Development Commission) (2006), *The Role of Nuclear Power in a Low Carbon Economy*, London: SDC.

SDC (2007), *Turning the Tide: Tidal Power in the UK*, London: SDC.

Seyfang, G., I. Lorenzoni and M. Nye (2007), *Personal Carbon Trading: Notional Concept or Workable Proposition? Exploring Theoretical, Ideological and Practical Underpinnings*, CSERGE Working Paper EDM 2007–03, Norwich: University of East Anglia.

Sorrell, S. (2003), 'Who owns the carbon? Interactions between the EU Emissions Trading Scheme and the UK Renewables Obligation and Energy Efficiency Commitment', *Energy and Environment* 14, 677–703.

Stern, N. (2006), *The Economics of Climate Change*, Cambridge: Cambridge University Press.

SustainAbility (2008), *Sustainability + Democracy*, Consultation developed by SustainAbility for the Environment Foundation, www.democracy.sustainability.com [14 March 2008].

The Royal Society (2002), *Economic Instruments for the Reduction of Carbon Dioxide Emissions*, London: The Royal Society.

Weale, A. (1992), *The New Politics of Pollution*, Manchester: Manchester University Press.

7

France: Towards an Alternative Climate Policy Template?

Joseph Szarka

Introduction

France was one of the first countries to adhere to the United Nations Framework Convention on Climate Change (UNFCCC), signing in 1992 and ratifying in 1994. Recognition of climate change as a real and imminent threat has been a constant of French policy-making ever since, reflecting the country's significant climate vulnerability. Temperature increases in mainland France are estimated to be 50 per cent higher than average global warming, prompting predictions of lower rainfall in summer and heat waves of comparable or greater severity than that of 2003 occurring more frequently (French Government 2006). Pursuant to article two of the UNFCCC, which states the objective of preventing 'dangerous anthropogenic interference with the climate system', the French government has committed to the view that a mean global temperature rise greater than 2 °C above the pre-industrial level constitutes 'dangerous climate change' (MEDD 2004: 69). In 2003, a 'factor four' reduction target – namely a 75 per cent cut in greenhouse gas emissions by 2050 – was announced by President Chirac (de Boissieu 2006: 13). This target was later incorporated into legislation in the 2005 Energy Bill.

Yet under the 1998 burden-sharing agreement, which programmed an aggregate eight per cent reduction in greenhouse gas emissions below 1990 levels for the EU for the 2008–2012 Kyoto commitment period, France agreed merely to stabilize emissions at 1990 levels. In practice, France is meeting this commitment, but only just. Hence a key question is whether a gap is opening between modest short-term outcomes and ambitious long-term aspirations. To illuminate the issues and outcomes, this chapter will firstly review climate policies in France and demonstrate how pragmatism and path dependency have shaped the French approach. Its second section will provide explanations for modest goals by cross-relating them to the 'standard recipe' for emissions reduction, whilst the third will explore proposals to develop an alternative climate policy template and thereby attain a 'factor four' target.

French climate policies in historical perspective

In its 1993 programme for combating climate change, France put forward a distinctive strategy for greenhouse gas reduction, arguing that the indicator for assessing liability should be emissions per capita. This approach still informs the French approach and bears similarities to the 'contraction and convergence' model promoted by Meyer (2000). Viewing the atmosphere as a 'global commons', Meyer's altruistic model allocates national emissions reductions using norms of international and intergenerational equity. It recommends a progressive transition to common levels of emissions by permitting an increase by developing nations whilst promoting deep cuts on the part of rich nations. In a comparable vein, the French Government (1997: 18) stated that:

> The allocation of reduction objectives among industrialized countries should be worked out to eventually lead to a convergence of emission levels based on appropriate indicators. France will strive to defend this principle in international negotiations underway and in the future, in order to ensure required fairness in the international allocation of the global effort to reduce greenhouse gases.

However, advocacy of the emissions per capita approach also reflects national interests. France has relatively low emissions per capita compared with other industrialized countries. In 1990, CO_2 emissions were 5.92 tonnes per capita in France as compared with an Organization for Economic Cooperation and Development (OECD) average of approximately 12 tonnes (Van Rensbergen *et al.* 1998). According to the European Environment Agency (EEA 2004), by the 2000s France had the third lowest emissions per capita in the EU-25 and the second lowest greenhouse gas emissions per unit of GDP.

The explanation for this lies in policy measures adopted subsequent to the 1970s oil price shocks. Principally, these were strict regulations and incentives to improve energy efficiency, high taxes on fuels, and diversification of fuel sourcing within the electricity generation sector (French Government 1995). A major expansion in hydroelectricity had also occurred in the 1960s, whilst in 1974 the Messmer government accelerated the nuclear power programme. By the 1990s, these sources generated 90 per cent of French electricity, reducing fossil fuel generation from 60 per cent of sourcing in 1973 to a residual ten per cent. Because hydroelectric and nuclear power are almost carbon-free at the point of generation, their expansion achieved large cuts in CO_2 emissions from the power sector, with national emissions falling by 26.5 per cent between 1980 and 1990, the biggest reduction in the EU (French Government 1995). In 1996, 378 terawatt hours of electricity were generated by French nuclear power stations, with savings of 117 million tonnes of CO_2 equivalent ($MtCO_2e$) compared with coal-fired generation or 59 $MtCO_2e$ for combined cycle gas turbines (French Government 1997). By the 2000s, electricity generation accounted for only eight per cent of greenhouse gas emissions in France, as compared with 36 per cent in Germany and 40 per cent in the USA (MEDD 2004). This positive but unplanned outcome offers a key instance of path dependency, in that the nuclear option was selected for reasons of energy security and economic competitiveness – and not environmental protection.

Yet the achievement of deep emission cuts ahead of other G7 nations left France economically vulnerable. In its first report to the UNFCCC, the French Government (1995: 6) asserted that 'the cost of new measures liable to be taken in France will often be higher than in the other countries of the European Union or the OECD'. Ironically, in becoming an 'inadvertent pioneer' (Szarka 2006) in the 1980s, France lessened its scope to make cheap cuts once the Kyoto regime was in place. In contrast, competitor nations such as Germany and the UK benefited from the arbitrary choice of the 1990 baseline due to 'wall-fall' emissions reductions in the former and the 'dash for gas' in the latter – yet still had greater potential for making low-cost emissions reductions due to specificities in their industrial structures. This configuration of unintended outcomes helps to explain particular constructions of national interests in climate policy.

In international discussions, France's negotiating strategy based on capping per capita emissions pressured the largest industrialized countries to accept the major burden of greenhouse gas reductions. But the various Conferences of the Parties – which led to the 1997 Kyoto Protocol and follow-on agreements – were characterized by hard bargaining, with all sides seeking to restrict their liabilities. The industrialized nations who signed the Protocol espoused softer emissions targets than originally mooted, with key emitters such as the USA and

Australia refusing to ratify at all. Yet the absence of targets for developing countries – including major emitters such as China and India – attested to a degree of implicit acceptance of equity norms benchmarked on emissions per capita. Indeed, article three of the UNFCCC set out the principle that 'the Parties should protect the climate system for the benefit of present and future generations of humankind, on the basis of equity and in accordance with their common but differentiated responsibilities and respective capabilities' (UNFCCC 1992: 4). During negotiations, however, recognition of 'common but differentiated responsibilities' fell short of explicit acceptance of emissions per capita convergence.

Thus, the French lost the argument on equity norms. France was also frustrated in advocating an international carbon tax. Initially, France gave conditional support to a carbon tax proposal provided that: (i) it was based solely on the carbon content of fuel (and not on energy content, which would have penalized nuclear power); (ii) it was implemented in all EU states; and (iii) precautions were taken to maintain competitiveness should other OECD states not implement equivalent measures (French Government 1995; IEA 1996; Godard 1997). The stress on uniform application of climate measures to ensure economic competitiveness has been a constant of the pragmatic French approach. However, with disquiet over the tax mounting in several member states (Zito 2000), at the 1992 Rio Summit the EU made the carbon tax dependent on establishment of comparable measures elsewhere in the OECD (Barrett 2003), a proposal rejected by the USA and Japan. Despite these set-backs at the international negotiating table, the French authorities held their line in intra-EU discussions. In the 1998 EU burden-sharing agreement, stabilization of emissions at the 1990 level – in contrast to the large cuts accepted by Germany and the UK – represented an acknowledgement of France's low emissions per capita (French Government 1997). But was this a soft target?

In the late 1990s, rapid economic growth pushed up emissions and an overshoot of some 25 per cent was predicted for the 2008–2012 Kyoto commitment period, leading the left-wing Jospin government to promulgate a national programme to fight climate change (Gouvernement français 2000). Three categories of preventive measures were envisaged: (i) voluntary agreements and emissions trading; (ii) a domestic carbon tax; and (iii) energy efficiency combined with recourse to renewable energy. In practice, each of these measures encountered their limits. In the late 1990s, voluntary agreements to cut emissions were made in the steel, aluminium, cement and glass-making sectors (French Government

2001). However, the carbon tax proposal (known as the *TGAP-Energie*) was struck down in December 2000. The French Constitutional Court ruled that the tax as proposed involved perverse discrimination between categories of energy consumer such that lower levels of consumption could incur higher levels of tax (Szarka 2003). Unlike the UK, France did not attempt a domestic experiment in emissions trading but waited for the EU scheme. Increased recourse to renewables was programmed by the year 2000 Electricity Act (discussed below). Nevertheless, energy efficiency remained the cornerstone of French climate policy.

In 2002, Jacques Chirac secured a second mandate as President, followed by a right-wing landslide in the parliamentary elections. The new Raffarin government criticized its predecessor's policies as inadequate, since the likelihood of an emissions overshoot by 2010 of 9.6 per cent remained (MEDD 2004). However, whereas Jospin's 'plural left' coalition (which included the French Green Party) had favoured new ecotaxes, the Raffarin government preferred other options. Ecotaxation had become associated with 'red-green' coalitions in France and Germany from which the right-wing wished to disassociate itself for reasons of political differentiation and pragmatism. The Raffarin government was receptive to industry's economic arguments against increased taxes, taking the view that 'we have to persuade rather than force companies, otherwise we risk undermining their competitiveness' (Baulinet 2002: 40). Thus, economic actors conditioned the government's perception of what was politically acceptable. The belated Climate Plan of 2004–2012 put forward some 60 additional proposals to reduce emissions, including incentives for buying low-emissions vehicles, support for biofuels, new regulations and energy efficiency certificates for buildings, wider use of energy labels and increased tax credits for efficient appliances (MEDD 2004). The use of the 'carrot', rather than the fiscal 'stick', points to the political sensitivities that inform the selection of climate measures in France. The plan projected 'savings' of 72 $MtCO_2e$ (MEDD 2004), whilst an update in 2006 aimed to 'save' a further six to eight $MtCO_2e$ (MEDD 2006a). If predictions prove correct, France will at least meet its emissions stabilization target for the 2008–2012 commitment period and, perhaps, achieve a cut of two to three per cent. This will be achieved with minimal use of the Kyoto flexibility mechanisms. Only four projects were registered by 2005 under the Clean Development Mechanism, consistent with France's position that mitigation should occur within national frontiers (French Government 2006). But it may also signal a preference to use flexible mechanisms as a fall-back option for the post-2012 commitment period.

Explanations of policy choices and outcomes

The principle of sectoral differentiation lies at the core of EU climate policy. It has produced a policy template which targets manufacturing industry and the energy sector (especially electricity generation) by (i) encouraging fuel switching away from carbon-rich sources (such as coal) to low-carbon (gas) or zero-carbon sources (renewables, nuclear); (ii) increasing energy efficiency; and (iii) industrial restructuring. The sources of this policy template were summarized by the EEA (2004: 12):

> The emission reductions in the early 1990s were largely a result of increasing efficiency in power and heating plants, the economic restructuring in the five new federal states in Germany, the liberalisation of the energy market and subsequent changes in the choice of fuel used in electricity production from oil and coal to gas in the United Kingdom and significant reductions in nitrous oxide emissions in the chemical industry in France, Germany and the United Kingdom.

The 'standard recipe' of targeting the largest industrial point sources offered the easiest and cheapest route to cut greenhouse gas emissions. It emerged during the allocation of reduction liabilities under the 1998 EU burden-sharing scheme and in the 2000s was translated into the EU emissions trading scheme (EU ETS). However, its impact in France has led to atypical outcomes.

During the run-up to the Kyoto Protocol, intra-European disagreements threatened the capacity of the EU to assume the mantle of climate leadership. In early 1997, the Dutch presidency of the European Council promoted the 'triptych' approach, which sought to allocate equitable shares of CO_2 reductions by taking into account energy supply systems and energy efficiency in industry (Andersen 2005). The 'triptych' approach divided national economies into three components: the electricity-generating sector; the energy-intensive, export-oriented sector; and the domestic sectors (Phylipsen *et al.* 1998). This allowed an objective and instrumental differentiation of emissions burdens on the basis of national industrial structures. It led to the establishment of the 'standard recipe' for emissions reductions across European states, with the largest cuts occurring in the electricity generation sector. The 'triptych' approach enabled a joint European position at Kyoto and led to the final burden-sharing agreement of 1998, with Germany and the UK pledging large cuts, Greece, Spain and Portugal securing big increases, and France arguing successfully for stabilization at the 1990 level on the basis of low per capita emissions and low emissions from the electricity-generating sector (Szarka 2008).

In a number of respects, the EU ETS represents an extension of the 'triptych' approach and of the 'standard recipe'. Unable to pioneer international carbon taxation, the EU fell back on sectoral measures applicable within its own frontiers, of which the EU ETS is now the centrepiece. As noted in Chapter 4, the scheme operates in two phases: a trial phase between 2005 and 2007 and an operational phase between 2008 and 2012 to accord with the Kyoto Protocol's first commitment period, and targets the largest industrial CO_2 emitters, in particular the electricity, energy, steel, cement, chalk, glass, ceramics, paper and cardboard sectors. Phase one covered some 12,000 installations producing 45 per cent of industrial CO_2 and equivalent to 35 per cent of aggregate greenhouse gas emissions in the EU (Andersen 2005). However, the total number of quotas allocated is dependent on the industrial structure of each member state. This revisiting of the 'triptych' approach produces comparable consequences at the nation-state level. The lion's share of emissions in Germany, Denmark and the UK comes from their electricity generation sectors, which therefore receive the most quotas.

The sectoral methodology underpinning EU climate policy-making has meshed tolerably well with the French policy style. The latter has been characterized in terms of *environmental meso-corporatism* (Szarka 2000), by which is meant the ring-fencing of a policy sector within which organized producer interests are entrusted with stewardship whilst being subject to the supervision of public institutions. Policy-making in longstanding environmental arenas such as industrial atmospheric emissions has been subtended by bilateral meso-corporatist bargains between industrialists and regulators (Szarka 2002). Although emissions trading was viewed with suspicion in France during the 1990s (Godard 2001), its implementation has fitted comfortably with national policy traditions, engendering institutional arrangements comparable with those found in pre-existing environmental policy arenas (Szarka 2006). The French authorities once again ring-fenced a meso-corporatist domain within which negotiation is undertaken between industry representatives and public officials. The National Allocation Plan (NAP) was drawn up by the ADEME (the Environment and Energy Efficiency Agency) and the register of emissions is kept by the *Caisse des Dépots et des Consignations* (a state-owned organization). Although producer interests exercise influence over the quantity and distribution of emissions quotas, the central state's authority is asserted by its right to set caps and apply sanctions as a last resort.

France has therefore demonstrated the institutional capacity to implement the EU ETS. Yet the scheme's impact is set to be low because the distribution of quotas mainly reflects existing national industrial structures, in particular the fuel mix in the electricity sector. At the EU level, 50 per cent of quotas go to the electricity sector, but only 25 per cent in the French case (Arnaud 2005), because 90 per cent of French electricity is sourced from (virtually) greenhouse gas free sources. In consequence, France is a marginal player within the EU ETS. In Phase one, Germany received 23 per cent of quotas, 11 per cent went apiece to Italy, Poland and the UK, but France held only seven per cent. Furthermore, whereas major corporations are the main recipients of quotas elsewhere in Europe, in France a large number of small establishments have been included – yet only 26 per cent of total CO_2 emissions were covered (Arnaud 2005).

For Phase two of the EU ETS, the French NAP of September 2006 proposed a ceiling on emissions of 151 $MtCO_2e$ per annum, as compared with verified emissions of 131 $MtCO_2e$ in 2005 (Actu-Environnement 2006). But with the European Commission refusing to tolerate lax proposals, a revised French plan was presented in December 2006, programming quotas of 128.86 $MtCO_2e$ per annum under 'grandfathered' emissions, plus an additional 3.94 $MtCO_2e$ set aside for market entrants (MEDD 2006b). This plan was approved in March 2007 and requires only a small improvement from established players. The French government did not exact more because its industrialists and small firms can argue that they would be economically disadvantaged in comparison with European competitors. Once again, the political obstacle to stronger measures is the influence of economic interests.

In summary, application of the 'standard' EU climate policy template to France is resulting in diminishing returns in the 2000s. The use of nuclear and hydroelectric power led to major greenhouse gas cuts from the electricity sector, whilst emissions from manufacturing industry fell by 22 per cent between 1990 and 2004 (MEDD 2006a). Thus, the stock of 'low hanging fruit' has largely been exhausted. In addition, the relevance of the domestic meso-corporatist style to new climate policy challenges is uneven, as will now be argued.

The prospects for future French climate policy

The French case shows that the premise according to which climate 'solutions' already exist and simply have to be implemented must be challenged. At one level, France is 'ahead of the curve' and has for some time been evolving an alternative climate policy template. Yet it clearly needs to go further. Moreover, the problems now faced by France will also become manifest in other developed countries once the 'standard recipe' is exhausted there too, although following suit, as noted earlier, would produce significant emissions reductions. The question is how a wider, more effective range of measures can be implemented within the strictures of existing institutional capacity but without arousing excessive political obstacles.

The structure of French emissions is set out in Table 7.1 in order to identify where the biggest problems lie. Since 1990, emissions have *fallen* in industrial processes by 27 per cent (15 $MtCO_2e$), in agriculture by nine per cent (10 $MtCO_2e$), and in waste by nine per cent (1.5 $MtCO_2e$) (French Government 2006). On the other hand, between 1990 and 2002 emissions *increased* by 23 per cent in the road transport sector (which accounts for a quarter of CO_2 releases) and by 8.8 per cent in the residential sector, despite substitution of coal by gas for heating. A similar upward trend exists in the tertiary sector. Although these sectors are contributing most to emissions growth in most industrialized

Table 7.1 Emissions calculations in the French 2004–2012 Climate Plan

Sectors	**$MtCO_2e$**			**$MtCO_2e$**	
	1990	**2002**	**2010 projection for business as usual**	**2010 percentage change relative to 1990**	**2010 projection with new measures**
Transport	121.5	149.5	175.1	44.10	154.8
Residential	89.5	97.4	116.6	30.30	99.9
Industry	141.2	115	118.3	–16.20	107.3
Energy	80.6	68.6	87.8	8.90	71
Agriculture and forests	116.1	108.6	108.1	–6.90	105.7
Waste	15.9	14.7	13	–18.20	12.5
Total France	**564.7**	**553.9**	**618.9**	**9.60**	**550.8**
Other state measures					–0.4
Carbon sinks					–3.2
JI, CDM					–1
Kyoto total					**546.6**

Source: MEDD (2004: 14 and 77).

nations, this dissection of trends by source categories provides pointers for policy formulation. France now has less 'low-hanging fruit' to pick in terms of CO_2 emissions from coal combustion and energy use in industry, or from nitrous oxide and methane sources in agriculture. Further reductions from those sources are, therefore, harder and more expensive to achieve than in other developed countries. In consequence, France must prioritize emission cuts from the tertiary, residential and transport sectors sooner and more substantially than neighbouring countries.

Prolonging or renewing the official scenario?

The French Government (2006: 58) stated categorically that 'energy efficiency is the main objective of French energy policy', specifying annual targets of a reduction of three per cent in greenhouse gas emissions and two per cent in end-use energy intensity up to 2015 (rising to 2.5 per cent by 2030) to occur primarily in the residential, tertiary and transport sectors. Policy measures included tax rebates to households of 40 per cent for the purchase of solar-heated water installations and 25 per cent for insulation, double-glazing and high-efficiency boilers. The energy efficiency labelling scheme (the A–F scale found on household appliances) is being extended to air-conditioning units, boilers, cars and even houses. A market will be created for 'energy savings certificates' designed to improve the thermal performance of buildings. Thus, the official scenario for emissions control departs little from France's long-standing energy policy framework and raises few political obstacles.

However, the results of the efficiency approach since 1990 have been productive but inadequate. The improvement in CO_2 intensity between 1990 and 2003 is an impressive fall of 17 per cent, approximating to 14 per cent per decade (MEDD 2006a). But in terms of aggregate emissions, the results are less encouraging. In 1990, total French greenhouse gas emissions were 564 $MtCO_2e$ and, by 2005, had fallen to 553 $MtCO_2e$, a decrease of 1.9 per cent over the period, or approximately 1.3 per cent per decade (Centre Interprofessionel Technique d'Etudes de la Pollution Atmosphérique 2007). In contrast, the 'factor four' target requires a reduction of 18 per cent per decade. It is hard to see how conventional efficiency savings can achieve this outcome and, thus, a step change in energy sourcing and consumption patterns will be required.

As regards energy supply, the official scenario continues to favour one of the main meso-corporatist bargains existing in France, namely the nuclear sector. The 2005 Energy Bill reinstated the French preference for nuclear power by the decision to build a demonstrator European Pressurized Reactor (EPR) plant in France by 2012. Former Industry Minister Nicole Fontaine (2006) outlined an ambitious schedule for the replacement of the domestic nuclear fleet, with a projection of 13 new reactors to be built by 2020 and another 24 by 2025. The French nuclear industry is also staking out an international leadership position. Anne Lauvergeon (2007), chief executive of Areva, the French nuclear supplier, stressed that markets are opening for nuclear power in China, India and the USA and set the target of capturing one-third of the global market. In November 2007, Areva secured an €8 billion contract to build two EPRs in China and supply them with uranium (World Nuclear News Organisation 2007). The French nuclear lobby, with support at the highest political levels, is exporting a long-standing domestic policy framework which links nuclear power, energy security and climate protection (Szarka 2008).

At the same time, signs of innovation can be detected in the increased emphasis on renewables. France is already the largest user of energy from renewables in Europe, producing 18.3 millions of tonnes of oil equivalent in 2004, amounting to 6.6 per cent of primary energy consumption, with 51 per cent coming from wood for heating, 31 per cent from hydroelectric power and 12 per cent from waste (French Government 2006). However, these are 'traditional' renewable energy sources. More diversification is needed into 'new' forms, such as wind and biofuels. To encourage diversification, European Directive 2001/77/EC set targets for electricity generation from renewables for each member state. For France, the 2010 target is 21 per cent (up from 15 per cent in 1997). Most of the increase is budgeted to come from wind power. The year 2000 Electricity Bill outlined a programme of expansion, with a 2001 decree establishing a Renewable Energy Feed-In Tariff (REFIT), guaranteeing kilowatt hour prices to targeted suppliers. Since the REFIT mechanism was a major cause of the dramatic growth of wind power in Denmark, Germany and Spain, similar consequences were expected in France (Szarka 2007a). In practice, expansion proved slower. At the start of 2007, Germany, Spain and Denmark had 20,622 megawatts (MW), 11,615 MW and 3136 MW of capacity, respectively, compared with 1958 MW in the UK and 1469 MW in France (Windpower Monthly 2007). Policy was reformed in 2006, with an increase in subsidy levels and the introduction of wind power development zones (Szarka 2007b). Yet doubt remains as to whether France can meet its 2010 target.

Energy crops were identified as a promising development to enhance energy security and reduce emissions. Moreover, their production does not challenge the traditional policy style. The French agricultural sector is highly corporatist, with an influential farmers' lobby. Reforms to the Common Agricultural Policy and increased global competition have provoked a crisis for French farming, so energy crops offer welcome opportunities for diversification (Bal 2005). Their production may yet renew traditional French goals of subsidizing agriculture and increasing national independence in energy supply. Proposals to expand biomass production were outlined as early as the first report to the UNFCCC (French Government 1995) and were implemented in the 2000–2006 Wood Energy Plan. Tax credits to households buying high-efficiency wood-burning stoves have stimulated expansion, and France has become the biggest producer of energy from wood in the EU (Delannoy 2007).

Progress in biofuels has, however, been faltering. EU Directive 2003/30/CE set the objective of substituting 5.75 per cent of vehicle fuels by biofuels by 2010. France brought this deadline forward to 2008, setting targets of seven per cent for 2010 and ten per cent for 2015 (MEDD 2006a). Yet acceleration proved faltering with output of biodiesel falling by 2.5 per cent between 2003 and 2004 to 348,000 tonnes, whereas it increased by 44 per cent in Germany to reach 1,035,000 tonnes (Anon 2005). New tax measures were introduced in 2005 to incentivize use, with plans to double production capacity of biodiesel to 1,800,000 tonnes per annum (French Government 2006). These measures will improve France's greenhouse gas balances, but the scale of the improvement depends on how fast fossil fuels are phased out.

The official scenario writ large?

The question of the speed and scale of reform was addressed in proposals put forward by Prévot (2007). Believing that concerns over the 'end of oil' confuse the real issue, Prévot argued that surplus quantities of fossil fuels remain available which, if combusted, will lead to a runaway greenhouse effect and global disaster. The aim must be to leave fossil fuels in the ground (and/or use carbon sequestration techniques which have yet to be mastered). However, achieving this aim would cause a collapse in energy prices to levels insufficient to sustain renewables. These premises led him to construct an interventionist model of energy and climate policy where the state reconfigures the energy sector. Prévot

(2007) offered three scenarios for a 30–40 year period, involving reductions in greenhouse gas emissions ranging from 33 to 60 per cent. The differences arose mainly from assumptions regarding (i) the scale of road transport increases (modelled to rise by either 15 or 35 per cent) and (ii) the scale of nuclear generating capacity (increasing from today's 63 gigawatts (GW) to either 93 GW or 133 GW). In his view, large-scale energy efficiency savings will make it possible for nuclear-sourced electricity – together with biomass – to heat buildings, fuel electric vehicles and power the industrial, tertiary and residential sectors. To incentivize these outcomes, the state would set prices through regulation, taxes and subsidies (Prévot 2007).

In certain respects, Prévot's vision is traditional French energy policy writ large. Up until the 1980s, France was renowned for an interventionist style of economic and industrial policy-making known as *colbertisme*, in which central government took entrepreneurial initiatives by providing investment capital and even setting up operating companies to implement so-called *grands projets*. Examples of these industrial 'grand designs' include telecommunications, aviation, high-speed trains and, of course, nuclear power. When Prévot (2007) proposed to make climate protection – and the reconfiguration of the energy sector – into the new *grand projet*, he explicitly situated himself within the interventionist French policy tradition. The latter concentrates exclusively on supply-side measures, a bias highlighted in Prévot's proposal to build dozens of nuclear power stations. Inevitably, such proposals provoke the objections of the 'Greens'. More fundamentally, they fail to address the demand-side reforms needed to achieve 'factor four' emission reductions.

Changing the cultures of consumption?

The defining characteristic of both the official scenario and Prévot's diagnosis is the bias to technocentrism. In other words, policy concentrates on improving the technologies of production. A complementary – and as yet underutilized – approach is to reform the cultures of consumption. However, encouraging changes in consumer behaviour requires a different order of policy-making to meso-corporatist, supply-side bargains. Modifications in industrial behaviour are predicated mainly on 'rational actor', profit-maximization cognitive frames, which predispose policy makers to the harnessing of market mechanisms. Clearly, all actors respond to economic stimuli, such as taxes and subsidies. But as Amadou *et al.* (2000: 3) note: 'further increases in energy tax levels may have limited impact on demand', due to limited options. Thus, with residential heating, consumers exercise a circumscribed choice between types of fuel that reflect the sunk costs of existing systems. With transport, choice of modes is limited or non-existent in the countryside – and France is a large, predominantly rural, country. In addition, consumer behaviour embodies a range of non-economic motivations, encompassing values, beliefs and goals which translate into 'lifestyle'. Yet governments have limited experience in steering 'lifestyle' change. In addition, political obstacles can arise from insecure public understanding and limited social acceptance (see Chapter 6 on the UK for further discussion of this issue).

Public opinion is slightly less concerned about climate change in France than elsewhere in the EU. In a Eurobarometer survey (2005), respondents were asked which five environmental issues they were most worried about: climate change topped the list at 47 per cent in the EU-15 but stood at 42 per cent in France, where larger proportions of respondents were most concerned with man-made disasters (55 per cent), air pollution (49 per cent) and water pollution (48 per cent). French governments have, however, increased their efforts to engage the public. In his 2002 presidential campaign, Jacques Chirac promised the energy debate that the French electorate had not seen in three decades of pro-nuclear policy. Over 2003, the Raffarin government duly organized a 'national energy debate', with distribution of information tracts and the holding of 100 public meetings and seminars, with the aim of feeding into new legislation (Leloup 2003; French Government 2006). A national publicity campaign to raise public awareness of global warming was also undertaken.

Initiatives in 'consciousness raising' have their merit, but tangible incentives need to be offered to low- and middle-income households to encourage investments in energy efficiency measures and fuel switching to renewables. Increased fiscal pressure or higher energy bills may prove counterproductive in encouraging changes in attitudes because of their regressive and inequitable effects. In the passenger transport sector, high excise taxes have helped to orientate purchases towards small diesel vehicles but have not stemmed major increases in private car ownership and aggregate distances travelled. Excise increases, especially in a period of rising oil prices, can provoke a violent public backlash, such as happened in 2001 when the French 'fuel escalator' had to be set aside. New solutions are required in transport, which has consistently been acknowledged as the sector in which emission growth is fastest (French Government 1995; 2006). Despite the development of high-speed intercity trains and new underground systems in major French cities, measures to encourage positive 'modal shift' – away from road and towards other forms of transportation – have produced modest results (Szarka 2004). The French Government (1995; 2001) was critical of EU liberalization of the road haulage sector because it encouraged modal shift to the detriment of the railways and inland waterways, arguing that increased competition led hauliers to compromise safety.

French road safety measures taken over 2003–2004 led to a decline in average speeds of vehicles and in fuel consumption, a development credited with achieving a two per cent fall in vehicle emissions (French Government 2006). However, in most years the increase in numbers of passenger vehicles has outstripped efficiency improvements. In the 1990s, France was already urging manufacturers to develop vehicles for urban use (rather than optimized for long-distance driving), but experiments with electric vehicles proved largely unsuccessful (Calef and Goble 2007). The 2003 Low-Emission Vehicles Plan offered support both to industrial research and to purchases of electrically powered vehicles (French Government 2006). Looking towards the next decade, renewed efforts are required to broaden the range of transport options, especially over short distances, and to encourage a widespread shift in mobility patterns. France's wealth of institutional experience in providing public transport solutions, its world-class vehicle industries and its agricultural potential for biofuel production provide essential preconditions for taking a pioneering role in revolutionizing passenger mobility. Ambitious and daunting as the transport revolution may seem, achieving the 'factor four' objective is impossible without it. This is an arena in which France can and should take bold initiatives.

Conclusions

Since the signing of the UNFCCC in 1992, French climate policy has displayed both consistency and specificity. Constant elements have been: the call for international convergence in per capita emissions; the high mitigation costs for France arising from early and deep greenhouse gas cuts; the preference for an EU regulatory framework (on carbon taxation, technology standards, behavioural rules, etc.) to ensure that climate policy does not impair France's economic competitiveness; and an energy policy which stresses energy efficiency and nuclear power. The result is a distinctive climate policy template, which France has sought to 'upload' to the EU and international levels. At the same time, France's Kyoto target is simply to achieve stabilization of its greenhouse gas emissions at the 1990 level – yet even this modest aim has proved challenging. Success in cutting industrial emissions has largely exhausted the 'standard recipe' for greenhouse gas cuts in the energy and manufacturing sectors, and has constrained the usefulness for France of the EU's flagship climate measure, the ETS.

The target of a 75 per cent cut in greenhouse gas emissions by 2050 will require France to switch from being an inadvertent to a deliberate pioneer. This will require supply-side improvements in the availability and quality of energy-sourcing alternatives, a wide range of energy conversion technologies (particularly in relation to renewable energies), as well as increasing efficiency and performance standards. But it also needs demand-led measures, with meaningful incentives for altered behaviour. France has the means to make this happen – in terms of natural resources, economic actors, technological expertise, institutional capacity and political traditions. But whereas the French policy style was traditionally based on environmental meso-corporatism (namely, a cooperative strategy between officials and industrialists), a different approach is required to engage the public and foster demand-led change. Investment in media campaigns to raise awareness of climate change has an important role. But reforming the cultures of consumption requires both changes in attitudes and the provision of practical alternatives. This in turn requires a step-change – in France, but also throughout the developed world – in which the current, narrow preoccupation with market-based and fiscal instruments gives way to ambitious public and private sector investments in the energy sources, technologies and lifestyles of the future. This will require radical change over the long term not just in the energy and manufacturing sectors, but also in urbanization patterns, construction, infrastructures and mobility. In brief, the 'factor four' target requires an alternative policy template whose beginnings we may be witnessing but whose ends have yet to materialize.

References

Actu-Environnement (2006), *Neuf états de l'Union Européenne se voient contraints de modifier leur plan national d'allocation de quotas*, http://www.actu-environnement.com/ae/news/2114.php4 [13 April 2007].

Amadou, T., P. Collas, J. Ellis and V. Matsarski (2000), *Report on the In-Depth Review of the Second National Communication of France*, http://unfccc.int/resource/docs/idr/fra02.htm [15 October 2007].

Andersen, M. S. (2005), 'Regulation or coordination: European climate policy between Scylla and Charybdis', in Hansjürgens, B. (ed.), *Emissions Trading for Climate Policy: US and European Perspectives*, Cambridge: Cambridge University Press, pp. 135–49.

Anon (2005), 'Le baromètre des biocarburants', *Systèmes solaires* 167 (mai–juin), 39–56.

Arnaud, E. (2005), 'Plan national d'allocations des quotas et territoires', *Revue de l'Energie* 564 (mars–avril), 92–9.

Bal, J.-L. (2005), 'Les multiple enjeux du développement des biocarburants en France', *Revue de l'énergie* 565 (mai–juin), 173–8.

Barrett, S. (2003), *Environment and Statecraft. The Strategy of Environmental Treaty-Making,* Oxford: Oxford University Press.

Baulinet, C. (2002), 'La lutte contre le changement climatique: les instruments de l'action gouvernementale et l'engagement des entreprises', *Réalités industrielles* (février), 39–44.

de Boissieu, C. (2006), *Division par quatre des émissions de gaz à effet de serre de la France à l'horizon 2050*, Paris: La Documentation française.

Calef, D. and R. Goble (2007), 'The allure of technology: how France and California promoted electric and hybrid vehicles to reduce urban air pollution', *Policy Science* 40, 1–34.

Centre Interprofessionel Technique d'Etudes de la Pollution Atmosphérique (2007), *Inventaire des émissions des polluants atmosphériques en France. Format SECTEN – mise à jour 2007*, http://www.citepa.org/publications/secten-fevrier per cent 202007.pdf [19 September 2007].

Delannoy, I. (2007), *Environnement: les candidats au banc d'essai*, Paris: Editions de la Martinière.

EEA (European Environment Agency) (2004), 'Greenhouse gas emission trends and projections in Europe 2004', Copenhagen: EEA report no. 5.

Eurobarometer (2005), *The Attitudes of European Citizens Towards the Environment*, http://ec.europa.eu/public_opinion/archives/ebs/ebs_217_en.pdf [6 December 2007].

Fontaine, N. (2006), 'Situation et perspectives de l'électricité nucléaire', *Regards sur l'actualité* 318 (février), 19–31.

French Government (1995), *First National Communication under the UN Framework Convention on Climate Change*, http://unfccc.int/resource/docs/natc/france1.pdf [15 October 2007].

French Government (1997), *Second National Communication under the UN Framework Convention on Climate Change*, http://unfccc.int/resource/docs/natc/france2.pdf [15 October 2007].

French Government (2001), *Third National Communication under the UN Framework Convention on Climate Change*, Paris: French Republic.

French Government (2006), *Fourth National Communication under the UN Framework Convention on Climate Change*, http://www.effet-de-serre.gouv.fr/images/documents/4th per cent 20National per cent 20Communication.pdf [19 September 2007].

Godard, O. (1997), 'Les enjeux des négotiations sur le climat', *Futuribles* 224 (octobre), 33–66.

Godard, O. (2001), 'Les permis négociables, une option crédible en France?' in Boyer, M., G. Herzlich and B. Maresca (eds), *L'Environnement, question sociale.*

Dix ans de recherches pour le Ministere de l'environnement, Paris: Odile Jacob, pp. 263–72.

Gouvernement français (2000), *Programme national de lutte contre le changement climatique*, Paris: République française.

IEA (International Energy Agency) (1996), *Climate Change Policy Initiatives*, Paris: OECD/IEA.

Lauvergeon, A. (2007), 'Le nucléaire, un des atouts maîtres dans la nouvelle donne énergétique mondiale', *Réalités industrielles* (février), 5–9.

Leloup, J. (2003), 'Le débat national sur les énergies', *Réalités industrielles* (août), 15–20.

MEDD (Ministère de l'écologie et du développement durable) (2004), *Plan Climat 2004–2012*, http://www.ecologie.gouv.fr [1 December 2004].

MEDD (2006a), *Actualisation 2006 du Plan Climat 2004–2012*, http://www.ecologie.gouv.fr/img/pdf/1er_doc_intro_plan_climat_final.pdf [16 November 2006].

MEDD (2006b), *Projet de plan national d'affectation des quotas d'émissions de gaz à effet de serre (PNAQ II). Période: 2008 à 2010*, http://www.consultationpubliquepnaq.com/download/PNAQII_v19ter.pdf [8 March 2007].

Meyer, A. (2000), *Contraction and Convergence. The Global Solution to Climate Change*, Totnes: Green Books.

Phylipsen, G., J. Bode, K. Blok, H. Merkus and B. Metz (1998), 'A triptych sectoral approach to burden-differentiation: GHG emissions in the European bubble', *Energy Policy* 26, 929–43.

Prévot, H. (2007), *Trop de pétrole: énergie fossile et réchauffement climatique*, Paris: Seuil.

Szarka, J. (2000), 'Environmental policy and neo-corporatism in France', *Environmental Politics* 9 (3), 89–108.

Szarka, J. (2002), *The Shaping of Environmental Policy in France*, Oxford: Berghahn.

Szarka, J. (2003), 'The politics of bounded innovation: "new" environmental policy instruments in France', in Jordan, A., R. Wurzel and A. Zito (eds), *'New' Instruments of Environmental Governance? National Experiences and Prospects*, London: Frank Cass, pp. 92–114.

Szarka, J. (2004), 'Sustainable development strategies in France: institutional settings, policy style and political discourse', *European Environment* 14, 16–29.

Szarka, J. (2006), 'From inadvertent to reluctant pioneer? Climate strategies and policy style in France', *Climate Policy* 5, 627–38.

Szarka, J. (2007a), *Wind Power in Europe: Politics, Business and Society*, Basingstoke: Palgrave Macmillan.

Szarka, J. (2007b), 'Why is there no wind rush in France?' *European Environment* 17, 321–33.

Szarka, J. (2008), 'Facing global climate risk: international negotiations, European policy measures and French policy style', in Maclean, M. and J. Szarka (eds), *France on the World Stage: Nation State Strategies in the Global Era*, Basingstoke: Palgrave Macmillan, pp. 181–98.

UNFCCC (1992), *United Nations Framework Convention on Climate Change*, Geneva: United Nations.

Van Rensbergen, J., L. Michaelis, A. Ghai, and T. Hadj-Sadok (1998), *Rapport de l'examen approfondi de la communication nationale de la France*, http://unfccc.int/resource/docs/idr/fra01.pdf [15 October 2007].

Windpower Monthly (2007), 'Operating wind power capacity', *Windpower Monthly* 23(4), 90.
World Nuclear News Organisation (2007), *Areva Lands World's Biggest Ever Nuclear Power Order*, http://www.world-nuclear-news.org/newNuclear/Areva_lands_world_s_biggest_ever_nuclear_power_order_261107.shtml [10 December 2007].
Zito, A. R. (2000), *Creating Environmental Policy in the European Union*, Basingstoke: Macmillan.

8
German Climate Policy Between Global Leadership and Muddling Through

Axel Michaelowa

Two decades of climate policy in Germany

Germany was one of the first countries in the world to embark on a process of defining concrete policy options to deal with climate change. The Parliamentary Enquiry Commission *Protection of the Earth's Atmosphere* was established in 1987, and its first report (Enquete-Kommission 1990) was a milestone in the analysis of potential climate policy measures, setting the tone for subsequent discussions in Germany. Climate policy has been an issue of considerable political importance in Germany ever since. German politicians have repeatedly proclaimed that Germany is a frontrunner in climate protection (Gabriel 2006) and German Chancellors have used this claim to stake out a leading role in international climate negotiations. Despite this, research on German climate policy is surprisingly sparse (exceptions include Michaelowa 2003; Bailey and Rupp 2004), although specific elements of German policy, such as negotiated agreements and ecological taxes, have attracted more attention (Bailey and Rupp 2005; Bailey 2007).

The purpose of this chapter is to explore whether German climate policy genuinely deserves the status of being a climate policy frontrunner, to identify a number of political obstacles to more ambitious climate policy in Germany and to suggest some political strategies that may assist in overcoming these barriers.

Institutional setting of climate policy in Germany

Responsibilities for climate policy within Germany are shared between several government ministries. The Interministerial Working Group on CO_2 Reduction (IMA) was set up in 1990 to coordinate climate policy and consists of several working groups, each dedicated to a specific issue area and led by the relevant line ministry. Energy supply is overseen by the Ministry of Economics and Technology (BMWi), transportation by the Ministry of Transport, Building and Urban Development (BMVBS), technology by the Ministry for Education and Research (BMBF), and agriculture and forestry by the Ministry for Consumer Protection, Food and Agriculture (BMVEL), while overall responsibility for climate policy *per se* rests with the Ministry for the Environment, Nature Protection and Nuclear Safety (BMU). A group on greenhouse gas inventories was added in October 2000 and a further portfolio covering Kyoto Protocol Joint Implementation and Clean Development Mechanism (CDM) projects was created in July 2005. The IMA published two policy papers in 1994 and 1997 in addition to issuing Germany's national programmes in 2000 and 2005 (Deutscher Bundestag 2000; 2005). The more recent 2007 climate policy programme was not based on IMA work, but rather on negotiations between the BMU and the BMWi led by the Chancellor's office.

This complex division of jurisdictions has created significant inter-departmental tensions, particularly between the BMU and the BMWi, which have hampered the coordination and coherence of German climate policy. Although climate policy is formally a BMU responsibility, the BMWi's coordination of energy policy means that different actor groups and considerations inform strategic decisions affecting energy structure, fuel sources and technology. The nature of decisions thus depends on the strength of alliances that each side can forge. The BMWi has repeatedly tried to reject policies that it found too constraining for industry (such as strict allowance allocations under the European Union emissions trading scheme (EU ETS)), ask for more cost-efficient measures (see BMWi 2004; 2007), or attack climate policy in general (Müller 2001). As there is no written long-term government energy strategy, energy policies are often framed in summit meetings between the Chancellor and industry representatives (see BMU 2006a).

Another important feature of German environmental politics is its tradition of consensus seeking and consultation, particularly with major employers and unions, which has encouraged a corporatist environmental decision-making style (Wurzel *et al.* 2003). On the international scene, German Chancellors have given high priority to climate policy, and within the EU have consistently played a key role in convincing less climate-policy-minded member states to accept EU policies. In international diplomacy, Germany has generally been faithful to joint EU negotiating positions.

Targets

German emissions targets have changed considerably over time, butwith a general tendency for strict target setting when climate policy becomes prominent in domestic political discussions, followed by a subsequent erosion of targets in small steps, often linked to developments in international climate policy (Table 8.1). For example, the 1987 enquiry commission proposed a CO_2 reduction target of 30 per cent below 1990 levels by 2005 and an 80 per cent reduction by 2050, which formed the basis of the federal government's first official target set in 1990. This target was quietly watered down in late 1990 and again in 1991, following German unification, by incorporating windfall emissions cuts in the former East Germany into existing targets, rather than increasing the targets. After minor changes in 1995 and 1997, further erosion of Germany's national targets occurred in 1998 as a result of the EU burden-sharing agreement (see Chapter 4), in order to avoid competitive disadvantages against other EU economies that had more lenient targets under the agreement. In 2002, the coalition Social Democrat-Green Party government specified a stronger new target for 2020, driven partly by the need to maintain the support of the Greens, which survived the change of government in 2005. The 2002 target was also a direct consequence of unprecedented flooding along the Elbe River, which created high public awareness of climate change and strengthened the electoral position of the Green Party in particular. Public awareness of climate change has remained high since (see FORSA 2005) and reached a highpoint in 2007, allowing few opportunities for the government to soften targets again.

Table 8.1 German emissions targets since 1990

Date of target setting	Base year	Target year	Gas covered	Reduction
June 1990 (West Germany)	1987	2005	Energy-related CO_2	25%
November 1990 (united Germany)	1987	2005	Energy-related CO_2	25% (West Germany) 'significantly more than 25%' (East Germany)
November 1991	1987	2005	Energy-related CO_2	25–30%
April 1995	1990	2005	Energy-related CO_2	25%
March 1997	1990	2008–12	CO_2, CH_4, N_2O	25%
June 1998	1990	2008–12	Kyoto basket	21%
October 2002	1990	2020	Kyoto basket	40%

Despite this, the fact that the government was able surreptitiously to bury the 25 per cent CO_2 reduction target for 2005 without provoking a public outcry does not bode well for other targets that are not enshrined in international agreements. The reasons for this are manifold but include changes of government, the rapid pace and fickleness of international climate politics, and concerns about the USA's withdrawal from the Kyoto Protocol.

The German climate policy mix over time

German climate policy is famous for its long lists of policy measures with widely varying characteristics. For instance, the 2000 programme listed 64 measures, and the 2007 one 29, some of which will actually increase greenhouse gas emissions, such as the operation of more powerful filter systems for nitrous oxide at thermal power plants. This again reflects the fragmented responsibilities for different aspects of climate policy along with the government's desire to tackle the issue on all fronts without over-regulating any one sector. Generally, the contribution of measures to emissions reduction is only calculated after they are agreed. For example, the BMU (2007a) calculates the effect of the programme set out in BMU (2007b) at 219 million tonnes of CO_2 ($MtCO_2$) reduction per year, based on bold assumptions about EU climate policy. An exception is the 2005 programme (Deutscher Bundestag 2005), which presented the reduction potential of each measure and a total reduction of ten $MtCO_2$ per year.

The main measures used in German climate policy are grouped thematically for further analysis. These are: ecological tax reform (ETR); emissions trading and project-based mechanisms (for example, the CDM); voluntary agreements; renewable energies; energy-efficiency measures; carbon capture and storage (CCS); non-CO_2 policies; and policies that lead to emissions increases.

Ecological tax reform

Although no cross-sectoral instrument for implementing climate policy existed in Germany for almost a decade, the late 1980s and early 1990s saw growing cross-party support for a unilateral ETR (Wurzel *et al.* 2003). However, a meeting between Chancellor Kohl, the chair of the Association of German Industry, and the chief executive of the chemical company BASF in 1994 led to the tax plan being dropped and substituted by a voluntary agreement (Wille 2003). Many within the ruling Christian Democrat Party, including the incumbent Environment Minister, Klaus Töpfer, backed the ETR, but lack of support from the Chancellor created an insurmountable veto point for the proposal. The concept of voluntary agreements was also more consistent with the German environmental policy style of close working with industry groups and regulated self-regulation, whereby the federal government proposes stringent regulation deliberately to force industry to offer a self commitment (*Selbstverpflichtung*) (Wurzel *et al.* 2003; Bailey 2007).

The ETR concept was only implemented following the victory of the Social Democrat-Green coalition in the 1998 federal elections. The reform consisted of five increases in taxes on mineral oil, gas and electricity, but is offset by reductions in employers' social security and pension contributions. However, the tax is effectively levied only on household energy use and transport fuels, as energy-intensive industries are largely exempted through a complex system of net burden rebates (Michaelowa 2003). Despite a campaign by the Christian Democrats in the 2005 election to abolish the tax, it was strongly supported by NGOs and was retained, although no further increases are planned, mainly because of developments in EU climate policy. Due to the tax and strong increases in vehicle fuel taxes over the preceding decade, transport fuel taxes in Germany are now among the highest in the world.

Emissions trading and project-based mechanisms

Emissions trading and similar market mechanisms played little role in the early stages of German climate policy. In contrast to EU member states like Denmark and the UK, domestic emissions trading was never seriously contemplated and when the Commission proposed an EU emissions trading scheme in 2001, it received a lukewarm reception from the German government. A Working Group on Emissions Trading to Combat the Greenhouse Effect (AGE) was set up under the lead of the BMU in 2000, but large industry groups opposed emissions trading in order to keep their voluntary agreements (Ströbele *et al.* 2002). Smaller industries were, however, more open towards the idea (Santarius and Ott 2002). Even in 2006, after the EU ETS had begun operating, the environment minister stated that some industry associations remained massively opposed to emissions trading (Gabriel 2006). However, after losing the battle over the introduction of the EU ETS, industry's focus switched to how it could benefit from the free allocation of emission allowances and schemes to provide financial rewards for early actions.

While the German government was among the first of the member states to submit its national allocation plans (NAPs) for the first and second trading periods, these were seen as insufficiently stringent by the Commission, leading to the embarrassing situation of Germany having to revise its plans on both occasions. The lack of stringency in the German NAPs was largely due to horse-trading between the BMU and the BMWi, whereby the BMWi accepted the continuation of a feed-in tariff system for renewable energies if BMU accepted an emissions cap that was about 15 $MtCO_2$ less stringent than the previous voluntary agreement. However, reflecting the ministerial divisions in German climate politics, it is conceivable that the BMU hoped the NAP would be rejected, as it was insufficiently influential within the German ministerial system to confront the BMWi directly. Moreover, a fuel-specific benchmark allowed coal-fired power plants to receive an allocation more than double that of gas power plants, and allocations for new fossil power plants were guaranteed for 14 years. Early actions, mainly in the former East Germany, where low-cost reductions could be achieved largely serendipitously through the closure of uncompetitive high-polluting facilities, also received additional allocations.

Despite warnings that the system would generate unfair windfall profits if allowances were allocated to companies that could increase product sales prices by the price of emissions allowances embodied in their product (particularly energy utilities, which have limited exposure to international competition), all allowances were allocated free during the first trading period. Due to increasing political debate about electricity producers receiving billions of euro windfall profits (Gabriel 2006), the allocation to utilities was slashed by 15 per cent in the second phase and many benefits from the previous NAP were taken away. Although auctioning was initially opposed even by the BMU (Gabriel 2006) due to a misunderstanding of the functioning of markets, pressure by parliament and carbon market lobby groups led eventually to nine per cent of allowances being auctioned (see BMU 2007c).

The government has continuously stressed that Germany does not need to import project-based emissions credits generated by the Kyoto mechanisms to reach its emissions targets (Gabriel 2006), and no fund or other vehicle for acquiring credits has been established. Despite this, the government used the discretion given by the Commission to the maximum possible when thresholds for importing emissions credits for participants in the EU ETS were announced. The initial proposal had been to restrict the use of Kyoto mechanisms to meet 12 per cent of Germany's allowance allocation. However, the government increased this to 22 per cent once the formula defined by the Commission allowed this. Currently, an initiative is being planned to support German companies to use Kyoto project mechanisms.

Voluntary agreements

Prior to the introduction of the ETR in 1999, voluntary agreements with industry – along with policies to support renewable energies – formed the mainstays of German climate policy. The first agreements were negotiated in 1995 and were made more stringent in 1996 by switching from relative to absolute targets and by including independent monitoring (Bailey 2007). The second round of agreements was finalized on a cross-sectoral level in 2000 but planned sector-level agreements were never confirmed, as the voluntary agreements were effectively overtaken by the EU ETS (Michaelowa 2003). In keeping with the German tradition of regulated self-regulation, a regulatory threat (in 1995, of a mandatory waste heat recovery ordinance and, in 2000, of compulsory energy auditing) was used on both occasions to secure the agreements.

Although the voluntary agreements were largely superseded in 2001 by the negotiation of the EU ETS, interestingly, allocations under the scheme were less stringent than the implicit allocations derived from the voluntary agreements. This is further explained by the fact that German voluntary agreements are rarely replaced by mandatory policies even if they are unsuccessful (despite the threat of regulation commonly being used to force agreements). However, the mandatory nature of the EU ETS caps had stronger implications for German industrial competitiveness (BMU 2007b). Despite these acknowledged failures, the 2007 climate programme envisages a future agreement with industry, coupling energy tax relief to energy efficiency efforts by companies (BMU 2007b).

Renewable energies

The oldest and most consistently followed climate policy instrument in Germany is the granting of feed-in tariffs to renewable energy producers, differentiated according to the technology used. While the laws granting tariffs have changed names over time, the principles of priority grid access for renewable electricity, as well as cost-covering tariff levels, have been upheld. This was due to clever lobbying by alliances that defended this system against attacks from the BMWi and the electricity utilities (Michaelowa 2005a). Recently, however, degression rates for tariffs (their decrease to successively lower rates) have been increased. The consequence of the tariff system has been a rapid growth in renewable electricity capacity, especially wind generation and photovoltaics, where Germany is a world leader in terms of installed capacity. Renewable electricity targets have repeatedly been raised (from 12.5 per cent of installed electricity capacity to be generated from renewables by 2010, set in 1997; 20 per cent by 2020, set in 2006; and 25–30 per cent by the same date, set in 2007), and generation tripled between 1990 and 2005. However, offshore wind generation has not flourished, while subsidies have increased from €0.3 billion in 1996 to €1.3 billion in 2000 and €5.8 billion in 2006 (BMU 2007d). Due to the high subsidy levels, overcapacity has been experienced in photovoltaics and wind turbine manufacturing for the last decade and emphasis has been placed on increasing production rapidly instead of increasing production efficiency. As a result, costs per installed megawatt (MW) have risen for both photovoltaics and wind since the early 2000s (Table 8.2).

More recently, liquid biofuels for transportation have benefited from fuel tax exemptions. The powerful farming lobby has pressed strongly for biofuel subsidies despite analyses by Henke *et al.* (2005) indicating that bioethanol production in Germany is neither competitive nor an efficient mitigation policy, and that biofuel imports are likely to increase. In the future, biogas feed-in gas pipelines are to be supported by granting priority and specific feed-in tariffs (BMU 2007b). Moreover, mandatory renewable energy-use targets for buildings (focusing on energy demand rather than supply management) are scheduled for development.

Energy efficiency measures

Energy efficiency measures have played a significant role in the German climate policy mix, but have repeatedly been sacrificed when they could

Table 8.2 Development of renewable electricity generation in Germany (TWh)

Type	1990	1995	2000	2006	Change (multiplier)	Tariff 2007 (Euro cents) per KWh
Hydroelectric	17.0	21.6	24.9	20.7	1.2	3.6–9.7
Waste	1.2	1.4	1.9	3.6	3.0	6.4–7.3
Biomass	0.2	0.7	2.3	14	63.0	8.0–21.0
Wind	<0.1	1.8	7.6	30.7	767	5.2–9.1
Photovoltaics	0	0	<0.1	2.2	2220	38.0–54.2
Total	18.4	26.5	37.7	71.2	3.9	10.9

Source: BMU (2007d: various pages).

be 'swapped' for other policy instruments. In 1995, for instance, the government axed plans for a waste heat ordinance when the first voluntary agreements were concluded. In 2000, regulations on the setup of cogeneration plants suffered a similar fate, while cogeneration subsidies agreed in 2002 have not stimulated significant uptake of this technology.

The only sector where energy efficiency has consistently been promoted is residential buildings, which benefit from interest rate subsidies for loans financing refurbishment. These were increased to €1.4 billion per year in 2007, while building energy efficiency standards are to be strengthened by 60 per cent in two steps in 2008 and 2012. In a bold move, obligations to improve energy efficiency during major retrofits are to be developed, albeit only if they meet economic viability criteria (taking subsidies into account). Financial penalties are also planned to improve compliance with building regulations and, if enforced, the regulations may have a major impact on German greenhouse gas emissions.

Carbon capture and storage (CCS)

Germany has a long history of coal and lignite production and use in electricity generation. The coal industry is currently undergoing a painful restructuring process, with mine closures and reductions in subsidies affecting economically vulnerable and politically sensitive areas in the Ruhr and the former East Germany. Many coal-fired power plants are also nearing the end of their economic lifetime, raising inevitable questions about replacement generation capacity and the need to capture and store emissions if new coal-fired power stations are commissioned. After an initially neutral stance on CCS technologies, the German government is now financing numerous research consortia and industrial players to develop pilot projects for CCS and sequestration (Michaelowa 2005b). There is a broad coalition of ministries, coal companies and research institutions (see BMWi, BMU and BMBF 2007) supporting the idea that, as a hi-tech country, Germany must not abandon coal technologies (Gabriel 2006). The government plans to develop binding standards to ensure captured CO_2 is sealed off permanently from the atmosphere and to guarantee secure and environmentally responsible long-term storage (BMU 2007b).

Non-CO_2 policies

Although the major focus of German climate policy is on reducing CO_2 emissions, other greenhouse gases have also been the subject of policy attention. Methane generation from waste is addressed by regulations to capture and use landfill gas, while industrial gases are covered by the voluntary agreement with the chemical industry to reduce nitrous oxide emissions by 47 per cent during its lifespan. Further reduction potential nevertheless remains, as shown by a substantial number of coal-mine methane and nitrous oxide reduction projects submitted under the Joint Implementation mechanism of the Kyoto Protocol. However, the coal-mine methane projects were rejected as they were found to be sufficiently subsidized by the feed-in tariff system.

Policies leading to emissions increases

German climate policy can only be fully understood by examining policies that are increasing, as well as reducing, greenhouse gas emissions.

The most high profile of these is the phase out of nuclear energy generation. Since the Chernobyl disaster in 1986, nuclear power has been strongly opposed by left-wing parties and environmental groups. This opposition peaked in 1998, when the Social Democrat–Green coalition brokered an agreement to phase out nuclear energy over a 25-year period. Despite intense and recurrent discussions, this decision has not been revoked so far and the first two plants have been decommissioned. Even with a change of government, policy makers argue that the loss of nuclear capacity can easily be covered by expansion in the renewables sector. Gabriel (2006), for example, compares the decommissioning of seven gigawatts (GW) of nuclear power with the commissioning of 19 GW of renewable energy capacity, although this overlooks the difference between reliable baseload power from nuclear power and potentially intermittent wind and solar power.

A second key issue has been the treatment of the car industry, which enjoys a special status in Germany illustrated by the fact that it has not been possible to introduce a general speed limit on German motorways and a lack of deterrents to the production of large luxury cars. Although relatively heavy fuel taxation creates an incentive for consumers to buy smaller cars, other demand-management measures such as differentiated car purchase or registration taxes, or the abolition of tax concessions for large company cars, have not been introduced. The German government has also taken active steps to avoid mandatory fuel efficiency standards at the EU level despite vehicle manufacturers' failure to reach voluntary EU-wide CO_2 emissions targets. This move failed, however, when the Commission adopted a proposal for legislation setting CO_2 performance standards for new passenger vehicles sold in the EU in December 2007 (European Commission 2007).

Where have emissions gone down?

In order to understand better the political obstacles to deeper cuts in German greenhouse gas emissions, it is instructive first to review where emissions have and have not declined. Overall, German emissions have reduced considerably and not only in the early 1990s when the emissions benefits from reunification were harvested. The 'hot air' from the collapse of heavy industry in the former German Democratic Republic only covers about 30 per cent of total CO_2 reductions (Schleich *et al.* 2001). Trends in German emissions also differ considerably according to sectors and types of greenhouse gases (see Table 8.3).

It is clear from this that non-CO_2 gases have reduced substantially. The chemical industry introduced thermal decomposition of nitrous oxide from adipic acid production (acids made from natural fats and used in nylon) in 1997, although emissions from nitric acid production have increased. Methane and perfluorocarbons (PFCs) have more than halved, the former through landfill gas recovery regulations, increased recycling and methane recovery from coal mines and wastewater, the latter through voluntary efforts of primary aluminium producers and semiconductor manufacturers. Sulphur hexafluoride reductions of 2.5 million tonnes of CO_2 equivalent ($MtCO_2e$) have been achieved in automobile tyres since the mid-1990s. Combined, non-CO_2 gases have achieved three percentage points of total emissions reductions in Germany.

Table 8.3 German emissions 1990–2005 (million tonnes CO_2 equivalent)

	1990	1995	2000	2005	Change 1990–2005
CO_2	1032	921	883	873	−15.4%
...of which transport	150	NA	172	152	+1.2%
Methane	99	81	65	48	−52.1%
Nitrous oxide	85	78	60	67	−21.6%
HFCs (base year 1995)	[4.4]	6.5	6.5	9.4	+44.6%
PFCs (base year 1995)	[2.7]	1.8	0.8	0.7	−59.0%
Sulphur hexafluoride (base year 1995)	[4.8]	7.2	5.4	4.7	−34.4%
Total	1232	1096	1020	1002	−18.7%

Source: Federal Environmental Agency (2007).

With regard to CO_2, energy and industry saw substantial reductions up to 1995, with a reversal from 2000 onwards. This was, however, offset by strong reductions in the transport sector, where an increase in emissions of 15 per cent from 1990 levels by 1999 was almost totally reversed by 2005. Household emissions have decreased by 20 per cent (Federal Environmental Agency 2007). BMU (2006b) calculates the combined effects of climate policy measures at 76 $MtCO_2e$.

Political framing and obstacles to German climate policy

A range of domestic and international factors have both contributed to and hindered the development of German climate policy. Cavender-Bares and Jäger (2001) identify several defining features of German environmental policy: its federal parliamentary democracy and tradition of strong state regulation; high public concerns about environmental issues; and the increased entwining of German and EU environmental policy (particularly via the EU ETS), which has challenged aspects of the German approach to climate policy.

Focusing for now on domestic climate politics, public pressure and the political doctrine of *Ordnungspolitik* have provided the general justifications for government intervention to create framework conditions for market actors, most notably through regulations and the adoption of the precautionary principle and best available technology as standard requirements in German law (Wurzel *et al.* 2003). Conversely, German environmental policy is biased towards incrementalism by ministerial divisions within the federal government, the sharing of some environmental competencies between the central and state governments, and Germany's bicameral parliamentary system, which requires major policy changes to receive the assent of both the *Bundestag* (the national chamber) and the *Bundesrat* (the state representative chamber). Two further components contributing towards incrementalism are the predominance of government by coalition, where support is needed from coalition partners for radical reforms (Cavender-Bares and Jäger 2001), and proportional representation, which has given the Green Party comparatively high access to the parliamentary process. Finally, the consensual and corporatist style of German climate policy (arising from its strong manufacturing base), coupled with policy traditions favouring voluntary agreements, has strengthened industry's ability to influence policy (Wurzel *et al.* 2003).

The general strategy chosen by German politicians to navigate these crosscurrents has been one that:

- Stresses Germany's role as a front runner in international climate policy by setting ambitious long-term targets, generating positive media coverage for the government;
- Exempts industry from burdensome measures;
- Focuses on high-visibility technologies whose costs are widely redistributed among the population. However, unduly strong rents for specific interest groups are not tolerated, leading to downward moves in feed-in-tariffs and auctioning of emissions allowances;
- Reacts strongly to meteorological extreme events, such as the Elbe river flood and winter storms.

The result is a strategy which has tended to promote high-cost measures. For example, relatively low-cost energy efficiency savings from industry have been delayed by successive governments' failure to create stronger financial incentives for industry to reduce emissions. The emissions reduction costs of photovoltaics are three orders of magnitude higher than for industrial energy efficiency and have not declined appreciably in the last decade. Studies and policy papers stressing the need for cost-efficient climate policies (BMWi 2004; McKinsey and Company 2007) have not changed policy makers' minds. McKinsey and Company (2007) find that, by 2020, German greenhouse gas emissions could be reduced by 26 per cent if all known options with mitigation costs up to €20/tCO_2e are implemented. A decrease of 31 per cent could be achieved if – while maintaining the nuclear phase out – the energy mix is adjusted to include a higher share of renewable energy, with costs of €32/tCO_2 for power generation from renewable energy sources and €175/tCO_2 for biofuels. Perplexingly, strategy elements that have enjoyed high success, such as the reduction of methane and industrial gases, rarely feature in political discourse.

Due to the consensus orientation of German politics, NGOs have integrated themselves strongly into policy networks to forge sometimes unlikely coalitions (Foljanty-Jost and Jacob 2004). This approach has enabled powerful alliances between interest groups favouring renewable energy technologies, environmental NGOs, and also farmers' associations and trade unions which stood to profit from land deals and increased employment (Michaelowa 2005a). Traditional industry lobbies, meanwhile, have tried to minimize the impact of climate policy wherever possible. In 2005, an environmental NGO unearthed a draft position paper by the Association of German Industry containing comments by members of the steel, coal and chemical associations showing their general opposition to the Kyoto Protocol and mandatory emissions targets (Bundesverband der Deutschen Industrie [BDI] 2005; Deutsche Umwelthilfe 2005). Further obstacles to more ambitious climate policy created by industry are its strong preference for voluntary agreements – despite these having been widely criticized for not going beyond business as usual (Ramesohl and Kristof 2001) – and its opposition to energy and fuel taxation (Sterner 2007). The allocation of emissions allowances in Germany under the EU ETS has yet to be analyzed in detail, although industry lobbying was a key factor behind the generous allocations set during the scheme's first stage.

In contrast, public opinion and media appear to be lesser impediments to climate policy in Germany than in many other countries. The German media has generally favoured stringent climate policy and has based much of its reporting on the scientific consensus embodied in the Intergovernmental Panel on Climate Change reports (Grundmann 2007). However, on occasions the media has presented more sceptical views, usually after a peak in attention to climate change. This happened most recently in mid-2007 after climate change had been at the forefront of television coverage in February and March 2007 (Polixea Portal 2007). Despite this, the general public has consistently supported climate policy and sceptics have not been able to mobilize significant public opposition.

Beyond the easy measures: future directions for German climate policy

From the evidence presented, Germany remains a giant on earthen feet in respect of climate policy and faces substantial challenges if deeper reductions in greenhouse gas emissions are to be achieved. The first key challenge is to shift from policies that are based on an array of measures with widely differing costs and a reluctance to use global carbon markets, and towards policies that clearly distinguish between short-term, cost-efficient options whose portfolio is optimized and the provision of incentives for long-term technological development. The second is how to challenge powerful industry groups without alienating them, while a third is to deal with political obstacles created by Germany's coalition-based and often factionalized system of climate governance.

In the short term, a decisive shift is needed from subsidies for costlier forms of renewable energy towards energy efficiency, whose costs would be capped by the price of emissions credits under the Kyoto mechanisms or their eventual successors. At first sight, this would appear to be straightforward, as it involves a shift from high to lower-cost options. Nevertheless, it also requires a political will to ignore lobbies from the costlier forms of renewable energy while finding formulae that force greater energy efficiencies from industry and other sectors while remaining acceptable to them. One option for doing this is by linking the generation of credits from project-based mechanisms, like the CDM, to the interests of German technology exporters (although concerns over the verifiability and additionality of some project-based credits may hinder public acceptance of this approach), or more generally by stressing the economic benefits of lower energy consumption. A bold policy measure would be for the state to offer to buy units of emissions reductions gained from energy efficiency measures at the current price of Certified Emissions Reductions from the CDM in order to create positive incentives for industry and technology developers. Such an approach is estimated to give a financial incentive of about one cent/KWh (kilowatt hour) saved. An even bolder approach would be to set targets for full greenhouse gas neutrality like that proposed by Norway, again using international carbon markets to make this acceptable to key interest groups. The scale and credibility of the CDM could be transformed if the German export industry took up this challenge and the government acted as an honest guarantor of the scheme.

A second strategy, focusing on long-term policy, would be to adopt the approach used by the Carbon Trust in the United Kingdom, where public seed money from energy taxes is used as venture capital for technology companies. Rather than current support for costly technologies, this would encourage competition between technology types and reinforce the potential cost-effectiveness of climate policy. For this strategy to be workable, clear 'sunset clauses' would be needed to stipulate the conditions under which financial aid would end once promising technologies are brought to market.

More costly but major emissions-saving technologies, such as CCS in new coal power plants, could be driven by mandatory regulation, similar to the regulation on sulphur dioxide scrubbers in the 1980s introduced as a reaction to 'forest dieback' across the country. Martinsen *et al.* (2007) see CCS as necessary for achieving reductions of more than 35 per cent but argue that its mobilization requires a carbon price of at least €30 per tonne, while Schumacher and Sands (2006) stress that CCS is unviable under a baseline climate policy scenario. Prescribing CCS or similar technologies would only make sense, however, in sectors where technological alternatives are unavailable, although high public acceptance of climate policy means that electoral opposition is less likely, even where consumers incur additional costs, provided the long-term benefits are clearly stated. Viebahn *et al.* (2007) nevertheless caution that CCS does not eliminate emissions but merely reduces them by 65–79 per cent. They also stress that renewables might develop faster and could be cheaper than CCS-based coal plants in the long term.

In relation to renewable energies, feed-in tariffs have gained considerable academic and political support as an instrument for expanding capacity (Lipp 2007; Walz 2007) but, at a political level, their rise was the result of a 'battle over institutions' where the parliament, informed and supported by an increasingly strong advocacy coalition, backed renewables policies and won against nuclear and coal interests (Jacobsson and Lauber 2006). This case illustrates the importance of cross-party and political–private-sector coalition building as a strategy for dealing with power and interest divisions between the German federal ministries involved in climate policy. Although the general direction of climate policy has not been politically contentious because contrarian viewpoints have not gained a strong foothold in public opinion, inter-departmental tensions are only likely to be resolved by nurturing pro-climate policy advocacy coalitions within and across key ministries. Achieving this is obviously likely to depend, first, on being able to persuade strong, economically minded ministries like the BMWi of the economic benefits of stronger policies, and, second, the active support of industry groups for specific initiatives that also benefit the German economy.

Further opportunities for reform arise from the continued and growing impact of EU climate policies on Germany. For various reasons, German governments have traditionally used the country's reputation as an environmental leader state to press for German policies to be adopted across the EU while reacting more warily (as in the case of the EU ETS) to EU initiatives that clashed with national preferences and policy traditions. As a result of this approach, Germany has sometimes adapted slowly to EU innovations, and the relaxation of its environmental leader image in interactions with the EU may assist German governments to ease away from entrenched policy approaches that gave the country early advantages but which have since hindered the country's progress beyond the 'easy' emissions reductions.

Conclusions

This chapter has argued that despite Germany's self-positioning as a global climate policy leader, it will face significant difficulties living up to this image in the future. Gabriel (2006) has reiterated that Germany should not sacrifice climate protection to short-term economic interests; however, the history of German climate policy shows that short-term economic interests have frequently won key political battles. This has led to a diverse basket of measures that often prioritized expensive policies and reinforced path dependencies. Change is now occurring with reforms such as the EU ETS, and in 2008 the environmental minister argued that the scheme was not designed to lead to a loss of companies or jobs in Germany. Consumers are, nevertheless, likely to incur increased costs as a result of such measures because cheaper measures within industry have so far been ruled out of discussions. The BMU has also tried to stress the financial, employment and market benefits of German industries being leaders in technological development (BMU 2006c); however, such messages are likely to be more convincing during periods of economic growth when industry and consumers are more prepared to shoulder the additional costs of climate policy. Political strategies are thus needed that break Germany away from its more entrenched policy traditions, including a clearer willingness to harness market forces to reach short-term emissions targets and the provision of support for promising technologies in a way that avoids creating path dependencies. This requires thinking beyond industry lobbies and towards advocacy coalitions, as has occurred in Scandinavian countries, the real pioneers of climate policy.

References

Bailey, I. (2007), 'Market environmentalism, new environmental policy instruments, and climate policy in the United Kingdom and Germany', *Annals of the Association of American Geographers* 97, 530–50.

Bailey, I. and S. Rupp (2004), 'Politics, industry and the regulation of industrial greenhouse-gas emissions in the UK and Germany', *European Environment* 14, 235–50.

Bailey, I. and S. Rupp (2005), 'Geography and climate policy: a comparative assessment of new environmental policy instruments in the UK and Germany', *Geoforum* 36, 387–401.

BDI (Bundesverband der Deutschen Industrie) (2005), 'Entwurf – Wettbewerbsfeld globaler Klimaschutz: deutsche Kernkompetenzen optimal nutzen', *Positionspapier, Stand 25.10.2005*, Cologne: BDI.

BMU (Bundesministerium für Umwelt, Naturschutz und Reaktorsicherheit) (2006a), *Results of Energy Summit*, http://www.bmu.de/files/erneuerbare_energien/downloads/application/pdf/energiegipfel_ergebnisse.pdf [20 October 2007].

BMU (2006b), *Demonstrable Progress Report: 2006 Report under the United Nations Framework Convention on Climate Change*, Berlin: BMU.

BMU (2006c), *Poster on Economic Benefits of Climate Protection*, http://www.bmu.de/files/pdfs/allgemein/application/pdf/plakat_klimaschutz01.pdf [20 October 2007].

BMU (2007a), *Effects of German 2007 Climate Policy Programme*, http://www.bmu.de/klimaschutz/downloads/doc/40259.php/ [10 November 2007].

BMU (2007b), *Integrated Energy and Climate Programme*, http://www.bmu.de/files/pdfs/allgemein/application/pdf/klimapaket_aug2007_en.pdf [10 October 2007].

BMU (2007c), *Legal Documents Regarding Allocation of EU Allowances for 2008–2012 in Germany*, http://www.bmu.de/emissionshandel/downloads/doc/39620.php [20 October 2007].

BMU (2007d), *Erfahrungsbericht 2007 zum Erneuerbare Energien-Gesetz*, Berlin: BMU.

BMWi (Bundesministerium für Wirtschaft und Technologie) (2004), *CO_2-Vermeidungskosten im Kraftwerksbereich, bei den erneuerbaren Energien sowie bei nachfrageseitigen Energieeffizienzmaßnahmen*, http://www.bmwi.de/BMWi/Redaktion/PDF/Publikationen/Studien/co2-vermeidungskosten-im-kraftwerksbereich-bei-den-erneuerbaren-energien,property=pdf,bereich=bmwi,sprache=de,rwb=true.pdf [30 October 2007].

BMWi (2007), *National Action Plan on Energy Efficiency*, http://www.bmwi.de/BMWi/ Redaktion/PDF/E/nationaler-energieeffizienz-aktionsplan,property=pdf,bereich=bmwi,sprache=de,rwb=true.pdf [30 October 2007].

BMWi, BMU and BMBF (2007), *Joint Declaration on CCS*, http://www.bmu.de/klimaschutz/downloads/doc/39998.php [30 October 2007].

Cavender-Bares, J. and J. Jäger (2001), 'Developing a precautionary approach: Global environmental risk management in Germany', in Clark, W., J. Jäger, J. van Eijndhoven and N. Dickson (eds), *Learning to Manage Environmental Risks: A Comparative History of Social Responses to Climate Change, Ozone Depletion, and Acid Rain*, Boston: MIT Press, pp. 61–91.

Deutscher Bundestag (2000), 'Nationales Klimaschutzprogramm. Fünfter Bericht der Interministeriellen Arbeitsgruppe CO_2-Reduktion', *Drucksache 14/4729*, Berlin: Deutscher Bundestag.

Deutscher Bundestag (2005), 'Nationales Klimaschutzprogramm. Sechster Bericht der Interministeriellen Arbeitsgruppe CO_2-Reduktion', *Drucksache 15/5931*, Berlin: Deutscher Bundestag.

Deutsche Umwelthilfe (2005), *Rauchzeichen aus der klimapolitischen Steinzeit*, www.duh.de/uploads/media/DUH-Hintergrund-BDI-Klimapapier_final_01.pdf [10 October 2007].

Enquete-Kommission (1990), *'Vorsorge zum Schutz der Erdatmosphäre'; des Deutschen Bundestages, Schutz der Erde. Eine Bestandsaufnahme mit Vorschlägen zu einer neuen Energiepolitik*, Bonn: Economica.

European Commission (2007), *Proposal for a Regulation of the European Parliament and the Council setting Emission Performance Standards for New Passenger Vehicles as part of the Community's Integrated Approach to reduce CO_2 Emissions*

from Light-Duty Vehicles, COM(2007) 856 final, Brussels: Commission of the European Communities.
Federal Environmental Agency (2007), *National Inventory Report 1990–2005*, Dessau: Federal Environment Agency.
Foljanty-Jost, G. and K. Jacob (2004), 'The climate change policy network in Germany', *European Environment* 14, 1–15.
FORSA (2005), *Poll on Public Views on Climate Policy*, http://www.bmu.de/files/pdfs/allgemein/application/pdf/klimaschutz_forsa.pdf [15 October 2007].
Gabriel, S. (2006), *Germany Remains at the Forefront of Climate Protection*, http://www.bmu.de/english/emissions_trading/current/doc/37401.php [20 October 2007].
Grundmann, R. (2007), 'Climate change and knowledge politics', *Environmental Politics* 16, 414–32.
Henke, J., G. Klepper and N. Schmitz (2005), 'Tax exemption for biofuels in Germany: Is bio-ethanol really an option for climate policy?' *Energy* 30, 2617–35.
Jacobsson, S. and V. Lauber (2006), 'The politics and policy of energy system transformation: Explaining the German diffusion of renewable energy technology', *Energy Policy* 34, 256–76.
Lipp, J. (2007), 'Lessons for effective renewable electricity policy from Denmark, Germany and the United Kingdom', *Energy Policy* 35, 5481–95.
Martinsen, D., J. Linssen, P. Markewitz and S. Vögele (2007), 'CCS: a future CO_2 mitigation option for Germany? A bottom-up approach', *Energy Policy* 35, 2110–20.
McKinsey and Company (2007), *Costs and Potentials of Greenhouse Gas Abatement in Germany*, Cologne: McKinsey and Company.
Michaelowa, A. (2003), 'Germany – a pioneer on earthen feet?' *Climate Policy* 3, 31–44.
Michaelowa, A. (2005a), 'The German wind energy lobby: How to promote costly technological change successfully', *European Environment* 15, 192–9.
Michaelowa, A. (2005b), 'Carbon capture and geological storage research – a capture by fossil fuel interests?' in Sugiyama, T. (ed.), *Governing Climate: The Struggle for a Global Framework beyond Kyoto*, Winnipeg: International Institute for Sustainable Development, pp. 69–75.
Müller, W. (2001), 'Wachstum – Energie – Klima: Wirtschaftspolitik im Spannungsfeld', *Speech at Centre for European Economic Research*, Mannheim, 29 June 2001.
Polixea Portal (2007), *Klimaschutz erneut Topthema der Fernsehnachrichten, Infomonitor März 2007*, www.polixea-portal.de/infomonitor [1 July 2007].
Ramesohl, S. and K. Kristof (2001), 'The Declaration of German Industry on Global Warming Prevention: A dynamic analysis of current performance and future prospects for development', *Journal of Cleaner Production* 9, 437–46.
Santarius, T. and H. Ott (2002), 'Meinungen in der deutschen Industrie zur Einführung eines Emissionshandels', Wuppertal Paper 122, Wuppertal: Wuppertal Institute.
Schleich, J., W. Eichhammer, U. Boede, F. Gagelmann, E. Jochem, B. Schlomann and H.-J. Ziesing (2001), 'Greenhouse gas reductions in Germany: lucky strike or hard work?' *Climate Policy* 1, 363–80.
Schumacher, K. and R. Sands (2006), 'Innovative energy technologies and climate policy in Germany', *Energy Policy* 34, 3929–41.

Sterner, T. (2007), 'Fuel taxes: an important instrument for climate policy', *Energy Policy* 35, 3194–202.

Ströbele, W., B. Hillebrand, A. Smajgl, E.-C. Meyer and J.-M. Beringer (2002), *Emissionshandel auf dem Prüfstand*, Hannover: Universität Hannover.

Viebahn, P., J. Nitsch, M. Fischedick, A. Esken, D. Schüwer, N. Supersberger, U. Zuberbühler and O. Edenhofer (2007), 'Comparison of carbon capture and storage with renewable energy technologies regarding structural, economic, and ecological aspects in Germany', *International Journal of Greenhouse Gas Control* 1, 121–33.

Walz, R. (2007), 'The role of regulation for sustainable infrastructure innovations: the case of wind energy', *International Journal of Public Policy* 2, 57–88.

Wille, J. (2003), 'Die Killer der Ökosteuer. Die Klima-Selbstverpflichtung', *Frankfurter Rundschau*, 12 November 2003.

Wurzel, R., A. Jordan, A. Zito and L. Brückner (2003), 'From high regulatory state to social and ecological market economy? "New" environmental policy instruments in Germany', in Jordan, A., R. Wurzel and A. Zito (eds), *'New' Instruments of Environmental Governance: National Experiences and Prospects*, London: Frank Cass, pp. 115–36.

9

Conflict and Consensus: The Swedish Model of Climate Politics

Lars Friberg

Introduction

In 2006 and 2007 Swedish climate policy was ranked the best in the world in an index developed by the non-governmental organization (NGO) *Germanwatch* (Germanwatch 2006). Despite this, Sweden, like all industrialized countries, is still in the early stages of developing a low-carbon society. This chapter describes the development of Swedish climate policy and its most significant policies, and provides independent analysis of the major obstacles to, and opportunities for, more ambitious emissions reductions. In general it will show that historical decisions, combined with aspects of the Swedish political system that facilitated the introduction of fiscal and regulatory policies, have led to comparatively significant reductions in emissions growth that have not been economically damaging. However, achieving rapid further reductions will require politicians to tackle politically sensitive issues related to lifestyle changes and, potentially, the role of nuclear energy. There are signs that this is happening, but the challenges ahead may test Sweden's coalition- and consensus-based welfare system.

Swedish greenhouse gas emissions

Swedish per capita energy consumption is relatively high compared with other OECD countries but its per capita greenhouse gas emissions are low (Regeringskansliet 2005). This high energy use is partly due to the rapid expansion of electric space heating during the 1980s, and the size of Sweden's large energy-intensive industrial sectors, especially pulp and paper, iron ore extraction and engineering. Since the oil crisis in the 1970s, however, Sweden has significantly de-coupled

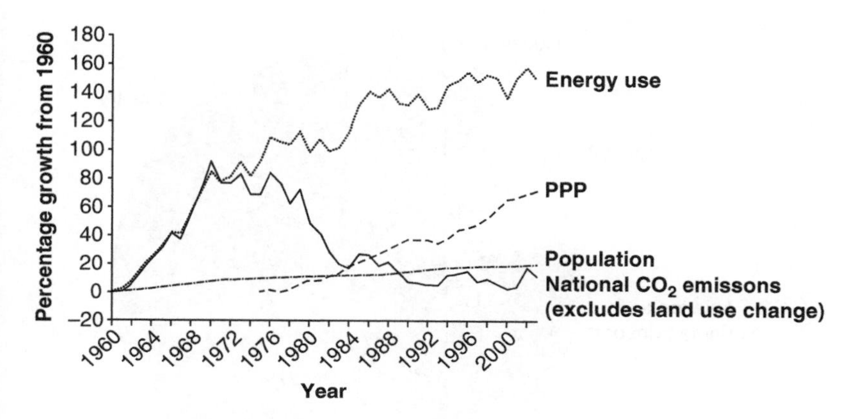

Figure 9.1 Key Swedish development statistics, 1960–2002
Note: PPP denotes Purchasing Power Parity, a measure of purchasing power that is not affected by currency exchange rates.
Source: Climate Analysis Indicators Tool (WRI 2007).

economic growth from emissions growth through a shift from oil- and coal-based space heating towards power generation based primarily on hydroelectric and nuclear power, each of which provides 40–45 per cent of Sweden's electricity (Figures 9.1 and 9.2). This makes the Swedish energy system one of the least carbon intensive of any OECD economy (Regeringskansliet 2005), but also increases the likely abatement costs of further greenhouse gas reductions.

Swedish greenhouse gas emissions also vary markedly from year to year, depending mainly on the severity of winters and rain and snow levels. When there is ample precipitation, hydroelectric power may prove sufficient even in cold winters. Otherwise electricity imports from higher CO_2 sources are required. The rate of economic growth, both generally and in key sectors, also has a significant impact on annual emissions, the last few years having seen strong economic growth and persistently high emissions.

Background to Swedish climate policy

When the *Riksdag*, the Swedish Parliament, first discussed climate change in 1988 it was decided that the government should ascertain the impact of energy consumption on atmospheric CO_2 concentrations and develop a programme aimed at ensuring that Swedish CO_2 emissions did not exceed 1988 levels (Regeringskansliet 2005). Following the

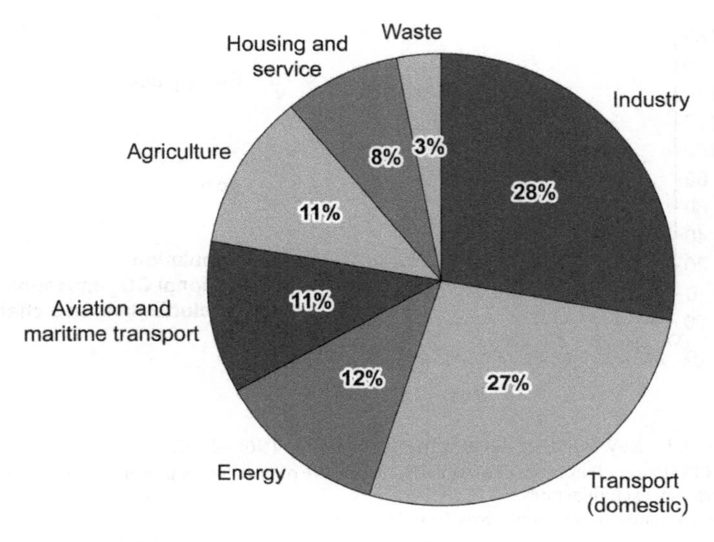

Figure 9.2 Swedish sectoral greenhouse gas emissions 2005
Note: Total emissions in 2005 were 72 $MtCO_2e$. Emissions from aviation and maritime transports are shown but not included in Swedish reporting to the UNFCCC, as they are not included in Kyoto Protocol reporting guidelines. Net emissions from land use, land use change and forestry are not included because they were negative in 2005.
Source: Environmental Protection Agency (Naturvårdsverket 2006).

release of the first Intergovernmental Panel on Climate Change (IPCC) assessment report in 1990, which provided the scientific basis for the United Nations Framework Convention on Climate Change (UNFCCC), the Swedish Parliament amended its 1988 climate policy objective to include limitations on emissions of all greenhouse gases, not just CO_2, in all sectors. The 1991 objective involved developing an action-oriented strategy for reducing climate impacts based on administrative and economic instruments and a commitment that Sweden, together with other Western European countries, should assume a proactive role in the international arena leading up to the signing of the UNFCCC in 1992 (Regeringskansliet 2005).

When discussing Swedish politics the dominant role of the Social Democrats must be noted. Sweden has had a Social Democratic government for 49 of the 63 years since Second World War. However many of these were minority governments that required the support of other parties, so consensus politics has been a major feature of the Swedish system. Thus climate policy, as with many other long-term policies, has been developed in cross-party parliamentary committees, and for this reason has at times been vulnerable to inter-party wrangling. Nevertheless, between October 1994 and October 2006 three consecutive Social Democratic governments shaped the majority of current policy. According to Bert Bolin, the first IPCC Chair and a member of the 2000 parliamentary committee, three political factions have been evident in Swedish climate policy (Stenman 2002). The Social Democrats, Left Party, Centre Party and Environmental Green Party have all favoured strong policy; the main conservative Moderate Party has argued against any measures that they felt might harm Swedish industry; and the Christian Democrats and Liberals accept that something has to be done but have also been concerned about economic impacts. These climate coalitions remained fairly consistent for many years but recently have started to change.

Sweden's progress on climate policy has also been substantially influenced by its entry into the European Union (EU) in 1995. Being a progressive country on environmental issues, Sweden advocated a strong EU position in the Kyoto Protocol negotiations in 1997. However, during discussions on the burden-sharing agreement (BSA) to apportion among member states the EU's joint Kyoto target to reduce emissions to eight per cent below 1990 levels by 2008–2012, Sweden secured a four per cent *increase* in its emissions above 1990 levels in recognition of the low carbon intensity of its economy (Gipperth 2007). Sweden is today one of few EU-15 countries on track to meet its Kyoto obligation under the BSA (EEA 2006). In addition, when the US Bush administration rejected the Kyoto Protocol in 2001, Sweden was holding the rotating Presidency of the EU, and Swedish Environment Minister Kjell Larsson and the EU Environment Commissioner Margot Wallström (a former Swedish Minister) both played crucial roles in maintaining EU backing for the Protocol despite the USA's decision to repudiate it.

Core components of Swedish climate policy

The present Swedish climate strategy is largely based on the climate policy decisions of 1993, 2001 and 2006 and the transport and energy policy guidelines drawn up in 1997 (Riksdagen 1993; Regeringen 1997, 2006; Klimatkommittén 2000). The 1993 climate policy decision was adopted as a national strategy for complying with the UNFCCC. Its goal was to stabilize CO_2 emissions from fossil-fuel combustion at 1990 levels by 2000 with reductions thereafter (Riksdagen 1993). The conservative government stressed that Sweden should avoid economically burdensome policies if competing economies did not implement similar policies, and advocated cost-efficiency. Economic instruments such as energy and carbon taxes have therefore played an important part in Swedish climate policy from the outset as cost-effective means of directing energy users away from fossil fuels.

In January 1999 an Environmental Code was introduced to bring together 15 existing environmental laws that had accumulated since the 1970s with the aim of promoting a coherent sustainable development strategy (Regeringskansliet 2005). The climate component of the code was translated into short-term and long-term targets. The short-term target was to reduce Swedish emissions of the six Kyoto greenhouse gases to at least four per cent *below* 1990 levels by 2008–2012, a substantially more ambitious domestic target than the four per cent *increase* contained in the EU BSA. Sweden's long-term target is based on the principle of emissions convergence across the world's population. Applying this principle, Swedish annual per capita greenhouse gas emissions should be reduced from its 2003 average of 7.9 tonnes of CO_2 equivalent (tCO_2e) to below 4.5 tCO_2e (Regeringskansliet 2005). This compares to the average per capita emissions in the USA of 19.9 tCO_2e and 3.5 tCO_2e in China, excluding emissions from land use, land-use change and forestry (WRI 2007).

Numerous programmes, initiatives, laws and regulations have been created to achieve these objectives and to promote the principles of sectoral integration and stakeholder involvement in the implementation of climate policy. The idea once again is that sectoral integration, together with fiscal steering mechanisms, will contribute to the cost-effective achievement of desired emissions reductions (Regeringskansliet 2005).

Sweden's most significant climate policies are focused in the areas of taxation, regulation, fiscal incentives and information efforts.

Taxation: the success and failure of the carbon tax

Energy and carbon taxes have played a key role in Sweden for many years, both as fiscal tax sources and as steering instruments. Other tax measures in this area include increased energy tax on fossil transport fuels; tax on methane emissions from landfills; and tax exemptions for biofuels. Energy taxes on fossil fuels, especially petrol but also other oil products, are comparatively high and, in theory at least, act as a powerful complement to the carbon tax while also having a significant revenue-raising function.

The energy tax system was reformed in 1991, based on the introduction of a new carbon tax and adjustments to energy taxes on fuels, the latter not being directly connected to the carbon content of fuels. In order to avoid excessive double taxation, general energy taxes were reduced by 50 per cent when the carbon tax was introduced. The system and tax levels have changed several times since 1991, but common features include lower or no taxes for heavy industry, farmers and electricity producers and higher tax levels for households and the service and retail sectors (Johansson 2000). The full tax rate has risen from 250 Swedish crowns in 1991 to 910 crowns per tonne of CO_2 (tCO_2) in 2005, the latter being roughly equivalent to €100 per tCO_2 (Regeringskansliet 2005). The clearest effect of the carbon tax is a fuel switch in the district-heating sector from oil to biomass (STEM 2006a; 2006b). In 1980 the total energy supply into the district-heating system was 34.5 terraWatt hour (TWh), of which 30.9 TWh (90 per cent) came from oil. By 2005 the share of oil in the system had reduced to 3.1 TWh (six per cent) while total energy supplied had risen to 54.8 TWh (STEM 2006a).

Regulation

Potentially the most important EU climate legislation is the European Emissions Trading Scheme (EU ETS). However, as Chapter 4 highlights, its potential is far from being fulfilled so far, with over-allocations of permits during the pilot phase and continued disputes over 2008–2012 national caps. Most assessments, and the European Commission's proposals for Phase 3, indicate that significant cuts in national allocations based on an EU-wide cap and expansion of the scheme to cover more sectors, including aviation, are necessary for the scheme to deliver major emissions cuts (Miljövårdsberedningen 2007). A significant early impact of the EU ETS in Sweden is that trading sectors (energy generators and selected major industrial sectors) became exempt from the Swedish CO_2 tax from 2006 (Johansson *et al.* 2005). Industry argued successfully that since their CO_2 emissions were now regulated by the EU ETS, paying the Swedish CO_2 tax constituted unfair double regulation (STEM 2006b).

Another significant regulatory measure is the 2003 Electricity Certificate System, which aimed to foster 10 TWh of new renewable electricity generation annually by 2015. In 2007 this target was raised to 30 TWh by 2020. Under this scheme renewable energy producers earn tradable 'green' certificates from the state regulator for each megawatt hour (MWh) they produce. The system does not have price controls or subsidies for renewable energy; instead the price for certificates is set by supply and demand in the market. Demand is created by obligatory quotas of renewable electricity for generators and users (EMI 2005).

The system was initially slow in creating additional renewable energy capacity, due partly to the reluctance of large energy utilities, such as Vattenfall, to invest. This behaviour is now starting to change. In November 2007 Vattenfall announced plans to construct 550 new wind turbines in the south of Sweden. If built, their 4 TWh output would equal the total capacity of the existing land-based Danish wind-energy sector. Unlike most existing turbines, these turbines are planned for forested areas, with 100 m tall towers and blades sweeping over the forest canopy (Johansson 2007). In addition, and as a reaction to high electricity prices, eight of Sweden's most energy intensive companies founded a joint wind power company aimed at providing themselves with cheap electricity.

Demand-side management and increased energy efficiency is another area in which Sweden has introduced new regulation. These include an environmental code promoting the efficient use of resources and energy recycling and reuse, and an energy-efficiency programme for manufacturing industries. Participating companies can avoid energy-use taxes on their electricity consumption by committing themselves to introduce an energy management system and measures to improve energy efficiency (Regeringskansliet 2005). Similarly, improvements in energy efficiency in the housing sector brought about by new building codes have roughly halved average energy consumption per square metre of living space since the 1970s. However these savings have been more than offset by increased energy consumption and an increase in total build space of almost 50 per cent over the same period (Regeringskansliet 2006). This has also been the case in the retail sector, where Sweden has the highest area of shopping centres per capita in the world (SR P1 2007).

In relation to non-CO_2 greenhouse gases, Sweden is implementing EU regulations phasing out a number of hydrofluorocarbons (Regulation EC/166/2006) as well as introducing national regulations to reduce methane emissions from agriculture (Regeringskansliet 2005). Producer Responsibility was introduced for a range of products in 1994, making producers and importers responsible for the collection, recycling, reuse or responsible disposal of waste products. This was followed in 2000 by a landfill tax and bans on combustible and organic waste being deposited in landfills. These have reduced the amount of household waste going to landfill by 70 per cent compared with 1993 (Sundqvist and Pollak 2007). Due to high household waste fees and the rapid expansion of the district-heating systems, waste has become a lucrative business in Sweden and waste management technologies have become one of Sweden's leading eco-technology exports.

Fiscal incentives: the local level

Local fiscal incentives are one of the hallmarks of Swedish climate policy. The most successful examples of this approach are the Local Investment Programmes (LIP), which operated for municipalities from 1996 until their replacement in 2003 by the Climate Investment Programmes (Klimp). These programmes have granted a total of six billion Swedish crowns (€660 million) from the state and have leveraged investments of almost 23 billion Swedish crowns (€2.5 billion) from local authorities, administrative regions and companies. More than half of Sweden's municipalities have participated, creating 307 investment programmes and 2569 projects in areas such as district-heating systems; anaerobic digestion of waste for biogas; support for transition to biofuels; energy efficiency measures; and local information campaigns about climate change. It is estimated that these have led to annual emissions reductions of 1.8 million tonnes of CO_2 equivalent ($MtCO_2e$) (Naturvårdsverket 2007).

Fiscal incentives: the failure to tackle transport emissions

Transport accounts for roughly 40 per cent of Swedish CO_2 emissions and shows the largest increase in CO_2 emissions since 1990 of any sector in the Swedish economy. The Swedish car market is significant in two main ways: it is dominated by 'national' brands and has the highest fleet fuel consumption rates in Europe (TT 2007a). Volvo and Saab, which were sold to Ford and GM respectively in the 1990s, make up over 40 per cent of the most sold cars in 2006 (Ekman 2007). Under the 1998 voluntary agreement between the European Commission and the European Automobile Manufacturers Association (ACEA), car makers collectively undertook to reduce average CO_2 emissions for the new car fleet to no more than 140 g/km by 2008. As the deadline approached it became apparent that the car industry, with a few exceptions, would fail to achieve this commitment, and in 2007 the Commission proposed binding limits on new passenger cars of a car fleet average of 120 g/km by 2012. Volvo's best selling model, the V70, currently emits 231 g/km (Ekman 2007). Apart from minor measures such as the introduction of tax breaks to make fuel-efficient or biofuel cars more affordable for companies and consumers, consecutive Swedish governments have failed to curb growing transport emissions due to the economic sensitivity of the car sector.

A prominent example of the political nature of transport policy is the introduction of congestion charges. Following the example of London, in 2005 Stockholm introduced congestion charging in the city centre combined with increased investment in public transport. This process produced political fallout for both major parties. In 2004 the Green Party made the introduction of congestion charging a requirement for their continued support of the Social Democratic minority government. This forced the government to overrule the Social Democratic controlled city council, which had promised voters before elections that no such system would be introduced. The system was introduced on a trial basis between August 2005 and July 2006, followed by a local referendum on its continuation. Evaluations of the trial concluded that the scheme had reduced congestion and improved public transport use and local air quality (Beser Hugosson *et al.* 2006). However the referendum, which was timed to coincide with national elections in 2006, was inconclusive: the city centre voted in favour of the scheme's continuation, whereas the surrounding (mostly conservative run) municipalities, which staged their own protest referenda, voted against. On the national level, the four-party coalition led by the conservative Moderate party won the national election for the first time since 1991. However, with climate change high on the political agenda, the new government faced pressure to show that it took the issue seriously and, against the wishes of local conservative politicians, decided to continue the congestion charge. One effect of the scheme is that sales of 'environmental cars' have boomed, as these models do not pay the congestion fees. Sales are also helped by a 10,000 crowns (€900) tax rebate on 'environmental cars' introduced by the government in spring 2006 (Hernadi 2007).

Information

Information campaigns to raise awareness of climate change have also been an important part of Swedish climate policy. While information alone is often insufficient to change behaviour, there remains a long-term need to build support and prepare society for more radical measures. One notable Swedish campaign is the *Klimatkampen,* a competition among high schools to design the most creative solution to reduce local emissions (IVL 2007). The decision in 2006 of *Aftonbladet,* Sweden's largest tabloid newspaper, to launch *Klimathotet* (the climate threat), a high-profile climate campaign with ample coverage, probably did more to raise awareness on the climate issue than any government campaign, especially among social groups that had not previously paid attention to the problem. Throughout 2007 climate change received massive media coverage, with many news portals creating permanent climate dossiers linked to their front page.

Extreme weather events have been another way in which climate change has gained increased public attention. In January 2005, storm Gudrun produced near-hurricane-force winds over south Sweden, felling or damaging 75 million cubic metres of forest, three times the total national annual logging volume. Electricity and communication infrastructure sustained damage from fallen trees and directly from the winds, and there were several fatalities. The Forest Agency regards the storm Gudrun as the worst ever to hit Sweden (Swedish Forestry Agency 2006).

Responding to Gudrun, parliament decided in June 2005 to create a climate and vulnerability inquiry committee to identify risks and formulate plans to adapt to climate change impacts. The work focused on infrastructure, estimating damage costs and adaptation strategies. The most serious impact identified was the risk from flooding if Vänern or other large lake systems breached their banks (Regeringskansliet 2007b). *Aftonbladet* published flooding scenarios, driving home general messages concerning the seriousness and immediacy of risks from climate-related weather events.

Sweden's international climate policy

In addition to formulating domestic policies and participating in the formulation of European policies, Sweden is also engaged in several multilateral climate initiatives. Mitigating and adapting to climate change is seen as an integral part of the overarching policy objective of poverty eradication. A third of Sweden's climate-related development assistance consists of multilateral support via mechanisms such as the Global Environment Facility (GEF 2005). Sweden also has some involvement with the flexible mechanisms of the Kyoto Protocol, which allow one country to finance emissions reductions in another country and count these as credits against its own emissions reduction target, through investments by the Energy Agency in four Joint Implementation projects in East Europe and in six Clean Development Mechanism projects in China, India and Brazil. Sweden has committed itself to generating 2.4 $MtCO_2e$ of emissions credits this way (ECON 2007), but the use of flexible mechanisms is so controversial in Sweden, especially among environmental groups (Eklöf 2006), that the government has decided not to use these credits towards Sweden's Kyoto target. Instead, they may be sold to other EU countries that are struggling to meet their Kyoto obligations.

Analysis of the politics of Swedish climate policy

Despite what looks like a successful consensus-driven policy, the process of elaborating Swedish climate policy has been fraught with political tensions. Within government, the Ministry of Environment generally argues for radical policy measures, while the Enterprise Ministry remains reluctant to adopt measures that could threaten the competitiveness of Swedish industry. In an effort to bridge this tension, the Social Democratic government reshuffled portfolios in 2004 to transfer housing and energy issues from the Enterprise Ministry to the Ministry of Environment, creating the world's second Ministry of Sustainable Development (the Canadian province of Québec created the first). However the reform remained superficial as the civil servants staffing the new ministry stayed the same and even kept their offices within the Enterprise Ministry. It also had little time to take root as the reform was overturned after the conservative alliance took office in 2006.

More generally, Sweden's tendency to elect coalition governments means that considerable bargaining is required to secure parliamentary support for new climate measures. The effects of this are difficult to assess precisely and can vary from government to government (for example, stronger measures are more likely when the Green Party is part of a coalition). Nevertheless, the need to obtain cross-party support can prove problematic for governments seeking to introduce measures that touch upon sensitive issues, like the car industry, or which are seen to be detrimental to the competitiveness of Swedish industry.

Industry

Industry was initially slow to react when the Swedish carbon tax was introduced in 1991, and it took several years to persuade the government that it should receive preferential treatment (Johansson 2000). The Energy Agency's own evaluation of the carbon tax finds that the differentiated tax system whereby industry and large energy users pay substantially lower taxes is partly the result of industry lobbying (STEM 2006b). Service and retail companies and households were not sufficiently organized as groups to counter the shift in the tax burden, which left them paying higher tax rates. In addition, electricity tends to be a relatively small budget item for these sectors, making it a less pressing issue for them. The exemption of industry and utilities from the carbon tax has nevertheless weakened this instrument, since these sectors comprise the largest point sources of emissions in Sweden and their long-term investment decisions will in large part shape Sweden's future emissions pathway. Success in reducing emissions in these areas thus depends heavily on the fortunes of the EU ETS and on political developments at the European rather than the domestic level.

Up until the EU's decision to adopt binding emissions targets for vehicle manufacturers, the car industry also successfully avoided painful policy measures. This can partly be explained by the focus of successive governments on cost efficiency, which is a near prerequisite for policy agreement given the frequency of minority Swedish governments. In essence, governments have been able to avoid tackling this politically sensitive sector as lower-cost reduction measures could usually be identified elsewhere. This lack of action can also be explained by a lack of political will to confront a car industry that is influential, vocal and an important source of employment. It is also telling that the Prime Minister of ten years, Göran Persson, started working as a lobbyist for the Swedish car industry within months of resigning in 2006.

As ever, however, the industry picture is a complex one, and a large number of Swedish companies are engaging positively with climate issues, including those in the car sector. Large corporations such as IKEA, Electrolux and Ericsson have all taken voluntary steps to reduce their climate impact, and large numbers of small- and medium-sized companies are exploring ways to reduce their carbon footprint or create market opportunities for other organizations. A significant and growing sector of Swedish industry also profits directly from emissions reduction: producers of biomass technology and different forms of green process design and system control all see growing exports in the expanding energy-efficiency and emission-management sectors. The question nevertheless remains whether industry groups are prepared to act and cooperate with government on a sufficient scale to achieve the large-scale emissions reduction recommended by climate science.

Environmentalists, sceptics and media

Swedish environmental NGOs such as the Swedish Society for Nature Conservation and the International Acid Rain Secretariat have criticized the Environmental Ministry for not doing enough on climate change and have argued for tougher targets for renewable energy, energy efficiency and modal shift in the transport sector. They have also played a significant role in informing citizens about how to reduce their climate impacts. As in the USA, Sweden also has its small share of vocal climate sceptic individuals and organizations, mainly from within neo-liberal think tanks like *Timbro* and *Captus* (Pamlin 1998). The editorial of the conservative daily newspaper *Svenska Dagbladet* was also for many years the main media articulator of climate sceptic views, although this never caught on in the other main newspaper editorials (Östlund 2005). This does not mean, however, that sceptics failed to influence political and public debates, as the media would often seek a 'balanced view' by pitting a sceptic against a mainstream scientist. However climate sceptics, never very influential, now appear to be a spent force, and the general impression is that the Swedish public is receptive to the ambitions of climate policy. The discussion has now shifted to how, not if, Sweden should rapidly reduce its emissions.

Nuclear: the continuing political headache

Although nuclear power became part of the Swedish energy-production mix in the 1960s, it has always been a highly contentious issue and, following an advisory referendum in 1980, the Parliament decided to decommission the 12 Swedish reactors by 2010. So far the only two reactors to be closed are those in Barsebäck in November 1999 and June 2005 in response to pressure from the Green Party and political frictions caused by their proximity to the Danish capital, Copenhagen. While not officially acknowledged, it is likely that the scheduled decommissioning of its nuclear capacity was a further reason for Sweden's relatively generous Kyoto target allocation under the EU BSA in 1998, since replacement energy sources are likely to be fossil-fuel derived. Nevertheless, utilities that own Sweden's remaining reactors, such as the German-owned E.ON Energie, have been loth to close them down because they are fully amortized and unaffected by the EU ETS, and so provide significant sources of revenue from generating almost half of Sweden's electricity.

The main stakeholders are divided on the nuclear issue, which also cuts through party allegiances. The main pro-nuclear groups are industry, the Moderate Party, the Christian Democrats and the Liberal Party. The Social Democrats oppose new nuclear reactors but a sizeable minority of its members, including some trade unions close to the party, argue in favour of continuing nuclear energy, while the main opposition groups are NGOs, the Greens and the Left Party. The Centre Party is perhaps in the most difficult situation, as its opposition to nuclear energy is unusual among Sweden's conservative parties and has caused difficulties in forming conservative coalitions. In order to make the current coalition work it was agreed that the nuclear issue would be taboo for the duration of its mandate, but this pact lasted little more than a year before the Liberal Party called for four new reactors to be built to tackle climate change (Björklund 2008). It should also be acknowledged that Sweden's radioactive waste problem is a quantitative and not a qualitative problem. Continued use will generate more waste but, as that problem already exists, decommissioning plants would not remove the problem. There is also the question of what energy sources could replace nuclear (Kärnavfallsrådet 2007).

Mobility: tampering with lifestyles

In addition to Swedes' fondness for large, domestically produced cars, foreign transport is another major contributor to emissions growth in Sweden. According to a recent study, ten per cent of Swedish emissions comes from tourism (Scott *et al.* 2007; TT 2007b), but curbing these emissions will require greater lifestyle changes than have so far been demanded and may lead to electoral repercussions for politicians who impose them. Although partly inconclusive, the referenda on the Stockholm congestion charge demonstrated that Swedish politicians cannot expect unequivocal support for behaviourally oriented measures. To avoid such constraints, carbon offset schemes are becoming increasingly popular among Swedish consumers and companies, while central government offices decided in November 2007 to start offsetting ministerial air travel with CDM carbon credits (Regeringskansliet 2007a).

Curbing Swedish road transport emissions may also be possible without great lifestyle sacrifices if appropriate messages are communicated. According to the Swedish Consumer Agency, if all car buyers bought the most fuel-efficient car model in the same size class without compromising security standards, car transport emissions would decrease by some 20 per cent, and if they shifted down one size class, emissions could decrease by 30 per cent (Konsumentverket 2007). Sweden is also a strong advocate of biofuels and is lobbying within the EU for the removal of EU import tariffs on ethanol as well as supporting research on second-generation cellulosic biofuels. However modal shift of transport is an area where Sweden has seen limited success: high-speed trains are competing with domestic aviation but road haulage tends to dominate the freight sector, due partly to the failure to create integrated networks for rail freight across borders.

Future politics of climate policy in Sweden

The year 2006 was in many ways a tipping point for the climate change issue in Sweden. The warmest autumn and winter in meteorological records resulted in intense media coverage. While Swedes might not mind warmer summers, the lack of winter and 'unnaturally' high temperatures on Christmas Eve brought home the message that something urgent needs to be done about climate change.

In July 2006 the Swedish Parliament adopted a long-awaited climate policy bill that reiterated the target of reducing emissions to at least four per cent below 1990 levels by 2008–2012 and reconfirmed Sweden's commitment to the EU target of keeping global warming to within 2 °C above pre-industrial levels. The bill also stated the aim that Sweden's emissions in 2020 should be 25 per cent below the 1990 baseline and that progress towards this target should be checked at regular so-called control stations (Regeringen 2006). Even when the four party Centre-Right Alliance ousted the Social Democratic party that had held power throughout the formative years of Swedish climate policy, the high public and political profile of climate change persuaded the new Prime Minister Fredrik Reinfeldt (Moderate Party), a politician not known for his environmental credentials, to confirm his government's commitment to existing targets and policies. Reinfeldt co-authored an opinion piece with his Environment Minister, Andreas Carlgren (Centre Party), expressing their support for a continued ambitious Swedish climate policy. In it they backed the 30 per cent greenhouse gas reduction target for 2020 proposed by the UK, Germany and France and re-stated Sweden's commitment to achieve a 25 per cent reduction in greenhouse gas emissions by 2020 (Reinfeldt and Carlgren 2006).

Exactly what policy mix Sweden (and the EU) will deploy to reach the 2020 targets remains uncertain but numerous organizations have begun to suggest strategies. The recommendation of the Scientific Council on Climate Issues is to continue to focus on domestic action, strengthen coordinated policies like the EU ETS and curb emissions from sectors such as transport and aviation. It also identified a potential need to use the Kyoto flexibility mechanisms to meet commitments (Miljövårdsberedningen 2007). Another strategy, advocated strongly by NGOs and the Greens, is continued shifting of the tax burden via reducing taxes on labour and increasing taxes on energy and natural resources, to make fiscal measures more palatable to industry and the public. A stronger role for green public procurement is another area of opportunity. Partly because of Sweden's strong welfare state tradition, the state is the single largest buyer of goods and services, so mandating a form of carbon rating for these purchases would support new technologies, reduce the public sector's carbon footprint and demonstrate the government's willingness to lead by example.

While nuclear energy is still controversial, attitudes towards it are slowly changing. Industry continues to lobby intensively in favour, and environmentalists are facing hard trade-offs as energy security and climate change dilemmas sink in. However, without a shift on the part of the Social Democrats to create a broad coalition in support of nuclear energy, it is unlikely that any new reactors will be built anytime soon. Upgrades of existing reactors to prolong their lifespan may be more feasible, but even if political barriers were removed it is doubtful whether private utilities would invest in new reactors given the huge upfront costs and the risk of public opposition.

In terms of political strategies to maintain the current tide of public support for more ambitious climate policies, politicians have the opportunity (via measures such as further reforms to the tax system) to take advantage of the Scandinavian welfare state ethos, whereby the public is generally accepting of state measures to tackle pressing issues like climate change provided that these are perceived as fair and responsibilities are distributed equitably across society. This is already evident in the fact that it has become politically untenable to criticize strong climate policy. The Moderate and Liberal parties are both developing political programmes on climate change, while the conservative Christian Democrats and Moderates, longstanding opponents of high fuel taxes and supporters of tax cuts in their 2006 election manifestos, have now abandoned this position and, despite high oil prices, now support higher petrol taxes (Wallberg 2007). How long this political zeal for progressive climate policies will survive remains to be seen but Swedish politics may have passed a threshold beyond which future discussion will be on how, not if, Sweden should curb greenhouse gas emissions.

The situation is probably somewhat different for industry. Although many industry groups now acknowledge the problem, they continue to oppose policies that they fear will impose competitive disadvantages. Some talk of relocation is undoubtedly scaremongering to sway the political process. Nevertheless, politicians will need to counteract this through a combination of appeals to public concerns about climate change (assuming that these do not waver if higher taxes are introduced) and the introduction of strong and coordinated EU policies to undermine competitiveness arguments. For a company to leave the world's largest single market is a much bigger step than leaving a country of nine million.

In the second half of 2009 Sweden will hold the EU presidency during a crucial juncture in international climate negotiations, as the international community prepares to finalize critical post-2012 UNFCCC negotiations on a successor to the Kyoto Protocol. Sweden is already involved in these preparatory efforts via a series of informal ministerial dialogues. After the first two meetings (Greenland 2005 and South Africa 2006) Sweden hosted the 'midnight sun dialogues' in the summer of 2007 in Lappland, where some of the more extreme effects of climate change are being felt. Such events have given Sweden opportunities to expand its political influence and utilize its tradition of trust-based politics as a means of overcoming some of the difficulties facing international climate policy.

Conclusion

Thanks to a combination of favourable circumstances and foresighted policies, the Swedish example has shown that it is possible to reduce greenhouse gas emissions while maintaining strong economic growth. Sweden is one of few industrialized countries on track to meet its Kyoto target and, according to recent projections, will also achieve its unilateral target of reducing emissions to four per cent below 1990 levels by 2010 (STEM 2007). Estimates of the aggregate effect of Swedish climate policies indicate that emissions in 2010 would have been 20 per cent higher if existing measures had not been implemented (Regeringskansliet 2005).

However, the reductions achieved so far have not demanded significant economic or lifestyle sacrifices. The likelihood is that achieving deeper emissions cuts will necessitate major changes in the way Swedes travel, consume and produce energy. Whether the Swedish political 'coalition and consensus' model can be sustained on climate change remains to be seen, especially if more painful measures are required. Nevertheless, Sweden has gained a head start on many other countries.

References

Beser Hugosson, M., A. Sjöberg and C. Byström (2006), *Facts and Results from the Stockholm Trial*, Stockholm: City of Stockholm Congestion Charge Secretariat.

Björklund, J. (2008), 'Sverige måste bygga fyra nya kärnkraftverk', *Dagens Nyheter*, 11 January 2008.

ECON (2007), *Miljöeffekten av JI och CDM Projekt*, Rapport 2007–07, http://www.naturvardsverket.se/upload/05_klimat_i_forandring/pdf/kontrollstation_2008/miljoeffekten_av_JI_och_CDM-projekt.pdf [10 November 2007].

EEA (European Environment Agency) (2006), *Annual European Community Greenhouse Gas Inventory 1990–2004 and Inventory Report 2006*, Copenhagen: European Environment Agency.

Eklöf, G. (2006), *Broken Illusions: CDM in Practice*, Stockholm: Swedish Society for Nature Conservation.

Ekman, I. (2007), 'Turmoil brewing in Sweden's love affair with big cars', *International Herald Tribune*, 4 July 2007.

EMI (Energimarknadsinspektionen) (2005), *Priser och Kostnader i Elcertifikatsystemet: Energimarknadsinspektionens Slutrapport i Projektet Transaktionskostnaderna i Elcertifikatsystemet*, Energimarknadsinspektionen, ER 2005:17.

GEF (Global Environment Facility) (2005), *Report on the Funding Status of the GEF Trust Fund as of September 30 2005*, www.gefweb.org [20 December 2006].

Germanwatch (2006), *Climate Change Performance Index*, http://www.germanwatch.org/ccpi.htm [20 December, 2006].

Gipperth, L. (2007), 'Sharing burdens in the European Union for the protection of the global climate: A Swedish perspective', in Lundqvist, L. and A. Biel (eds), *From Kyoto to the Town Hall: Making International and National Climate Policy Work at the Local Level*, London: Earthscan.

Hernadi, A. (2007), 'Dieselbilar nådde rekordnivåer', *Svenska Dagbladet*, 2 January 2007.

IVL (Swedish Environmental Research Institute) (2007), *Klimatkampen*, http://www.klimatkampen.se/ [5 December 2007].

Johansson, B. (2000), *Economic Instruments in Practice 1: Carbon Tax in Sweden*, Stockholm: STEM, Swedish Environmental Protection Agency.

Johansson, B., M. Björsell and M. Normand (2005), *Effekten På Utsläpp Av Förändrade Styrmedel Införda Efter 2004*, Naturvårdsverket och Energimyndigheten: Underlag inför den andra nationella fördelningsplanen för utsläppsrätter för perioden 2008–2012.

Johansson, H. (2007), 'Största satsningen på vindkraft', *Dagens Nyheter*, 16 November 2007.

Kärnavfallsrådet (Swedish National Council for Nuclear Waste) (2007), *Nuclear Waste in Sweden*, Stockholm: Kärnavfallsrådet.

Klimatkommittén (2000), *Förslag till Svensk Klimatstrategi*, Stockholm: Statens Offentliga Utredningar.

Konsumentverket (2007), *Stora Kommunvisa Skillnader i Personbilars Utsläpp*, Pressmedelande 2007-05-15, www.konsumentverket.se [4 June 2007].

Miljövårdsberedningen (2007), *Vetenskapligt Underlag för Klimatpolitiken*, Vetenskapliga rådet för klimatfrågor, Miljövårdsberedningens rapport 2007:03.

Naturvårdsverket (2006), *Sweden's National Inventory Report 2007: Rapport till Klimatkonventionen UNFCCC*, Stockholm: Naturvårdsverket.

Naturvårdsverket (2007), *KLIMP Klimatinvesteringsprogrammet*, http://klimp.naturvardsverket.se/mir/ [October 20, 2007].

Östlund, C. (2005), *Klimatfrågan i Dagspressen – En Analys Av Den Bild Svenska Dagstidningar Förmedlar Genom Sina Ledarartiklar När Det Gäller Klimatförändringarna*, Örebro: Örebro University.

Pamlin, D. (1998), 'Lobbying i en globaliserad värld', in Ahlsson, G., J. H. Bergström, A. Sundström and D. Pamlin (eds), *Lobbying*, Stockholm: Statens Offentliga Utredningar, p. 146.

Regeringen (1997), *Energipolitiska Riktlinjer*, Prop. 1996/97:84, Riksdagen.

Regeringen (2006), *Nationell Klimatpolitik i Global Samverkan*, Prop. 2005/06:172, Riksdagen.

Regeringskansliet (2005), *Sveriges Rapport om Påvisbara Framsteg*, Ds 2005:57, Fritzes förlag.

Regeringskansliet (2006), *På Väg Mot Ett Oljefritt Sverige, Kommissionen Mot Oljeberoende*, http://www.regeringen.se/content/1/c6/07/57/91/b9f3b671.pdf [5 December 2007].

Regeringskansliet (2007a), *Regeringen Klimatkompenserar för Flygresor*, http://www.regeringen.se/sb/d/9739/a/92587 [21 November 2007].

Regeringskansliet (2007b), *Sverige Inför Klimatförändringarna – Hot och Möjligheter*, Miljödepartementet, http://www.regeringen.se/sb/d/8704/a/89334 [5 December 2007].

Reinfeldt, F. and A. Carlgren (2006), 'Utsläppen av växthusgaser ska minska med 30 procent', *Dagens Nyheter*, 18 December 2006.

Riksdagen (1993), *Riksdagens Klimatpolitiska Beslut*, 1992/93:179, Stockholm: Riksdagen.

Scott, D., S. Gössling and C. de Freitas (2007), 'Climate preferences of tourists: evidence from Canada, New Zealand and Sweden', 3rd International Workshop on Climate, Tourism and Recreation, 19–22 September 2007, Greece: Alexandroupolis.

SR P1 (Sveriges Radio P1) (2007), *Efter Köpfesten*, in Stenholm, S. (ed.), *Godmorgon Världen*, 28 October 2007 Stockholm: Sveriges Radio P1.

STEM (Swedish Energy Agency) (2006a), *Energy in Sweden: Facts and Figures*, ER 2006:44, Stockholm: STEM – Swedish Energy Agency.

STEM (2006b), *Styrmedlens Interaktion: En Analys Av Hur Sex Ekonomiska Styrmedel Bidrar Till Klimatmålet Och Till Försörjningstrygghet*, ER 2006:37, Stockholm: STEM – Swedish Energy Agency.

STEM (2007), *Prognoser För Utsläpp Och Upptag Av Växthusgaser*, ER 2007:27, Stockholm: STEM – Swedish Energy Agency.

Stenman, D. (2002), *Science in Swedish Climate Policy: Vetenskaplig Förankring i Svensk Klimatpolicy*, Linköping: Linköpings University.

Sundqvist, J.-O. and F. Pollak (2007), *Utvärdering Av Svensk Avfallspolitik i Ett Systemperspektiv*, Malmö: Avfall Sverige.

Swedish Forestry Agency (2006), *After Gudrun: Lessons Learnt Following the Storm in 2005 and Recommendations for the Future*, www.skogsstyrelsen.se/stormanalys [23 January 2008].

TT (The Local) (2007a), 'Svenska bilar äldst i EU', *Dagens Nyheter*, 10 September 2007.

TT (2007b), 'Svensk turism en stor klimatbov', *Dagens Nyheter*, 12 November 2007.

Wallberg, P. (2007), 'Alla partier är för dyrare bensin', *Dagens Nyheter*, 3 December 2007.

WRI (World Resources Institute) (2007), *Climate Analysis Indicators Tool (CAIT) Version 4.0*, http://cait.wri.org World Resources Institute [4 December 2007].

10
Prometheanism and the Greek Energy *Zugzwang*

Iosif Botetzagias

Introduction

Greece is bound by the climate policies and targets of the European Union (EU) but succeeded, after tough negotiations, in securing a 25 per cent increase in its greenhouse gas emissions from 1990 levels by 2008–2012 as part of the EU's burden-sharing agreement, rather than a decrease in emissions. In terms of domestic policies to reach this target, the main strategy employed by Greek governments has been to move away from reliance on domestically available and cheap lignite towards natural gas and renewable energies for power generation. Other measures such as nuclear energy and energy taxes were never seriously considered due to widespread public opposition, and generally Greek governments have shied away from more aggressive approaches.

However, the expansion of natural gas and renewable energy generation has been seriously delayed by grass-roots opposition to renewable energy projects, while the phase out of lignite and the privatization of the Public Power Corporation (PPC), Greece's dominant energy producer, have met with strong resistance from trade unions amidst news reports of substantially increased electricity prices. It is now widely accepted that Greece will miss its 2008–2012 emissions target and face even more serious obstacles in the post-Kyoto period.

Following this introduction, the next two sections detail the main developments in Greek climate policy over the last 15 years and other policy options available. This is followed by a review of the shortcomings of current strategies. The conclusion then highlights the rather Promethean character of Greek climate-change discourse and suggests options for further emissions reductions.

Greek climate policy 1990–2006

Over the past 15 years, Greece has introduced numerous policies to reduce its greenhouse gas emissions. These are contained in Greece's four National Communications to the UN Framework Convention on Climate Change (UNFCCC) (hereafter: FNC (1995), First National Communication; SNC (1997), Second National Communication; TNC (2003), Third National Communication; and FONC (2006), Fourth National Communication), and two national Action Plans for Climate Change (1995 and 2002). The main idea running through these documents is the modernization and rationalization of the Greek energy system (Table 10.1). The FNC (1995) advocated supply-side adjustments promoting natural gas, renewable energy sources and cogeneration systems, increased efficiency in conventional power generation and the use of cleaner technologies. On the demand side, similarly modest proposals were advanced: more rational energy use and energy conservation in all sectors. The implementation of these measures was to be supported 'either by *administrative policies* which focus on ... regulations, or by *economic policies* which strive to modify the behaviour of the "players" involved, and the criteria according to which their energy-related decisions are adopted' (FNC 1995: 12). Similar rationalization calls were made in the SNC (1997) and, even for the period after 2000, the SNC highlighted energy efficiency, energy conservation, natural gas and renewable energies as its main objectives.

Greece ratified the Kyoto Protocol in 2002 (Law 3017/2002) and, in the same year, adopted its Second Action Programme for Climate Change (SAPCC 2002), with the goal of limiting greenhouse gas emissions increases to 25 per cent above 1990 levels by 2008–2012. These recommendations were then reproduced, and similarly modest policies to those in the SNC proposed, in the TNC (TNC 2003). The years between the TNC and the FONC (2003–2006) saw a change in government in March 2004, while the early 2000s were also marked by the adoption of the EU emissions trading directive (see Chapter 4, also Damro and Méndez 2003; Cass 2005). Greece, almost a year late and after the opening of European Commission infringement procedures (Buchner *et al.* 2007), was the last member state to submit – after virtually no public consultation – its National Allocation Plan (NAP). The plan allocated 71.1 million tonnes of CO_2 equivalent ($MtCO_2e$) of allowances free of charge per year to 141 businesses for 2005–2007. The PPC gained 74 per cent of these allowances but still needed to spend €12.6 million and €10.1 million in 2005 and 2006, respectively, to acquire additional permits (PPC 2007).

Table 10.1 Greek policies and targets to curb emissions (EU target for 2008–2012, +25 per cent compared to 1990 base year)

Referred to in	Main policy measures	Situation (from 1990)	GHG emissions projections with:	
			No measures	**Measures and additional measures**
FNC (1995)	Penetration of natural gas in all sectors including cogeneration		**in 2000**: +26.8%	**in 2000**: +15% (±3%)
SNC (1997)	Promotion of renewable energy sources for electricity generation and heat production		As above	
SAPCC (2002)		2000: +23.4%		**in 2010**: *With measures* : +35.8% *With additional measures* : +24.5%
TNC (2003)	Energy conservation in the industrial and residential–tertiary sectors			
FONC (2006)	Promotion of energy-efficient appliances/equipment in the residential–tertiary sectors	2000: +18.5%		**in 2010**: *With measures* : +34.7% *With additional measures* : +24.9%
	Structural changes in agriculture and the chemical industry	2003: +23.3%		
	Transport and waste-management options			

Greece's second NAP covering 2008–2012 was submitted for a ten-day public consultation in June 2006. Though the draft plan envisaged an allocation of 344.5 MtCO_2e per year, the plan actually submitted in September 2006 proposed a figure of 346.7 MtCO_2e per year. The Commission conditionally approved the Greek plan in November 2006, but requested an annual reduction of 6.4 MtCO_2e. In spring 2006, the FONC to the UNFCCC was submitted (FONC 2006). This included no new measures beyond those specified in the SAPCC, except for references to the Kyoto flexibility mechanisms and the financing mechanisms contained in the Third Community Support Framework. The Communication also noted that the Greek government 'has not committed any resources that could be used for the acquisition of emissions credits' (FONC 2006: 96), though several studies on the economic effects of using Joint Implementation and the Clean Development Mechanism to achieve Greece's Kyoto target have been financed by the Ministry for the Environment.

In summary, both Greek action plans and all National Communications have had a strong focus on supply-side measures and electricity production which are in effect monopolized by PPC. To an extent, this is unsurprising given PPC's heavy reliance on lignite-based energy production (Table 10.2). To an extent, this plan worked well. In 1999, natural gas accounted for just 0.4 per cent of Greek electricity production, but by 2006 it exceeded 20 per cent (DESMIE (Hellenic Transmission System Operator) 2006). In terms of the impact of these plans, however, latest data indicate that Greek emissions rose by 23.4 per cent above 1990 levels by 2003 (FONC 2006). Yet, this overall increase masks diverging trends, ranging from a reduction in nitrous oxide emissions to an increase of over 250 per cent in 'F-gases' caused by rising demand for

Table 10.2 Lignite energy/electricity contribution: energy greenhouse gas emissions

	1990	**1995**	**2000**	**2005 (projection)**
Indigenous Energy Production (million tonnes oil equivalent, Mtoe) of which, % lignite	8.77 81.2	8.99 83.5	9.99 82.3	10.76 81.3
Indigenous Electricity Output (Mtoe) of which, % lignite	2.99 72.4	3.54 69.6	4.59 64.2	5.66 51.4
Energy-related GHG emissions [all fuels] (Mt CO_2e)	74.7	73.9	76.8	78.2 (2003 data)

Sources: IEA (1998:107–9); IEA (2002:113–15); FONC (2006:58, Table 2.4).

residential and vehicle air-conditioning (FONC 2006). These highly differentiated trends are due to the Greek national plans' emphasis on CO_2, which accounts for almost 80 per cent of Greek greenhouse gas releases, and the lack of a clear strategy for other greenhouse gases (International Energy Agency (IEA) 2006) apart from the transposition of EU directives and regulations concerning landfilling and ozone-depleting substances (FONC 2006).

The plans also envisaged that the transition to natural gas and renewable energies would be supported mainly by regulatory and economic instruments focused on direct investments and subsidies. Financing was made available through a number of operational programmes co-financed by the EU, and several laws promoting renewable electricity sources. Most importantly, Law 3468/2006 provided for the compulsory acquisition by DESMIE of all electricity produced from renewable sources, at prices well above those charged by PPC to electricity users, for a minimum ten-year period. The difference is currently covered by a feed-in tariff on electricity bills, now standing at €0.4 per MWh.

Any other measures? Energy taxes and nuclear energy

Energy taxes

The IEA report on the SNC, published in 1998, stressed the importance of properly applying the 'polluter pays' principle or 'internalising external costs' (IEA 1998: 112). However, no mention of either principle was made in the FNC (1995) or SNC (1997). The First Action Plan for Climate Change (FAPCC 1995) nevertheless entertained the introduction of a coal tax to lower the costs of introducing natural gas to the PPC. A CO_2 tax was also discussed to secure more financial resources to implement the plan, yet various tax levels were found to have a smaller impact on CO_2 reduction than 'technological measures both on the demand and supply sides, even when the financing of the necessary interventions is to be assumed by the State' (FAPCC 1995: 33–4). In the early 1990s, the EU also debated proposals for a common carbon/energy tax, starting at $3 per barrel of oil, rising to $10 by 2000. Both proposals were blocked by a coalition of the UK and cohesion countries (Spain, Portugal, Ireland and Greece) for reasons of national sovereignty and the economic impacts of the tax (see Chapter 4, also Ringius 1999a; Padilla and Roca 2004).

Similar ideas were contained in the SAPCC (2002); however, Rapanos and Polemis (2005) concluded that taxes alone would fail to curb

Greece's CO_2 emissions to their target level by 2010, and that taxes needed to be complemented by further regulations and increases in renewable energy, although a CO_2 tax may be a necessary *extra* measure. Thus, the FONC (2006: 123) notes that 'The results of the "with measures" scenario clearly show that, despite the substantial changes that have already been realized or have been adopted in this sector, energy associated GHG [greenhouse gas] emissions continue to increase.' In the ensuing sensitivity analysis of energy sector emissions (FONC 2006), a $41.8 per tonne CO_2 tax was the only measure found to curb energy sector emissions significantly.

Energy/carbon taxes have, thus, been a recurring theme in the rhetoric of Greek Action Plans and Communications, but have failed to be implemented because of their potential repercussions for the existing Greek energy framework and, in particular (i) Greece's heavy dependence on lignite for energy production; (ii) the state policy of cheap electricity provision; and (iii) the Aegean islands' dependence on oil-based energy production. At the time of the IEA report, lignite accounted for roughly 80 per cent of Greek domestic energy production and 70 per cent of electricity supply (IEA 1998). Consequently, Greek CO_2 emissions per unit of primary energy supply were the highest of any OECD country in 1996. Nevertheless, the abundance of state-owned and PPC-utilized lignite meant that Greek industry and household electricity prices were below the European IEA member countries' weighted average and had, in fact, fallen in real terms since 1987. Household prices reduced by 30 per cent between 1987 and 1996, also encouraged by macro-economic policies designed to control inflation (IEA 1998). Between 1990–2003, residential electricity consumption increased by 81 per cent (IEA 2006), yet prices continued to fall to reach the lowest for household consumption in Purchasing Power Standards in the EU-27 by 2007 (Eurostat 2007). Obviously, the inclusion of environmental externalities into lignite-related energy production would result in substantial increases in end-user electricity bills.

A different yet equally distorted situation has arisen in the Greek islands, most of which are not connected to the mainland electricity grid and rely on fuel-oil and diesel generation. In 2000, this accounted for around eight per cent of national electricity consumption (IEA 2002). Despite the higher costs of energy generation on the islands, due to reasons of social and economic cohesion, a uniform tariff system has been created (IEA 2002) and, again, forcing islanders (who comprise ten per cent of the Greek population) to pay the full environmental costs of their energy would lead to major price increases.

Public opinion data on fiscal measures to curb emissions reveal a more complex picture. Early opinion polls indicated that Greeks viewed increased energy taxes *coupled with* an equal decrease in other taxes positively, although this figure fell from 65.4 per cent in 1993 to 49.9 per cent in 1996. The corresponding EU averages were 55 per cent and 46.1 per cent, respectively (Eurobarometer 1997). Yet this has had more to do with other pollution problems. In 1996, only nine per cent of Greeks thought that the most important energy challenge over the next decade was to reduce the greenhouse effect (the second lowest score in the EU, average 17.5 per cent), compared with cutting pollution (52.4 per cent) and price stability (25.9 per cent) (Eurobarometer 1997). Opinion polls in 2006 also showed only 35 per cent support for paying more for energy from renewable sources compared with other sources, down from 42 per cent in 2002 (EU average: 34 per cent and 37 per cent, respectively) (Eurobarometer 2002; 2006).

Nuclear energy

Nuclear energy has never been seriously considered in Greece. Greece is a non-nuclear country and a proposal in the mid-1970s to site a nuclear reactor close to Athens met with huge demonstrations (Botetzagias 2001). The issue has never resurfaced and the first Greek Action Plan on Climate Change (FAPCC) (1995: 11) mentions 'Greece's negative position on the issue of nuclear energy' in relation to CO_2 abatement strategies. In terms of public opinion, the percentage of Greeks who thought nuclear energy posed a low pollution risk dropped from 22 per cent in 1986 to five per cent in 1991 (the lowest score within the EC) (Eurobarometer 1991). Similarly, support for nuclear power stations fell to 6.1 per cent, the lowest in the EU (down from 15 per cent in 1989), while those feeling that it posed an unacceptable risk increased to 68 per cent (the highest in the EU) (Eurobarometer 1997).

Accounting for the Greek politics of climate change

Fisher (2003) argues that, alongside national concerns and key actors, Greece's position on climate politics has been particularly strongly shaped by the country's membership of the EU. Greece, as with other southern EU members, is generally considered to be one of the EU's environmental 'laggard' states (Börzel 2003), with frequent 'foot-dragging' to contain attempts by other member states 'to upload their domestic policies to the European level' (Börzel 2005: 170). This has been based on fears that stricter environmental regulations would place a disproportionate burden on Greece's developing industrial sector and employment. The cohesion countries have, thus, frequently sought to link acceptance of higher environmental standards to economic subsidies, exemptions and/or reductions in their national targets (Börzel 2005).

From the outset of EU climate policy in the early 1990s, Greece and the other cohesion countries pressed for lighter measures than their wealthier northern European counterparts (Ringius 1999b). This concept of common but differentiated responsibilities was given expression in the 1998 EU burden-sharing agreement, where Greece secured its 25 per cent increase in greenhouse gas emissions in recognition of its economic needs, rapidly rising emissions (35 per cent between 1985 and 1995) and the cost of stronger measures (European Environmental Agency 1996). This fuelled fears by other member states that the 2004 accession states would ally themselves with existing laggard states to block more ambitious climate policies (Skjärseth and Wettestad 2007). The same researchers and interviews conducted for this chapter nevertheless provide little evidence that these countries were forming a consistent veto bloc on climate policy (Costa 2006).

In general, the Greek style of environmental policy has been characterized as reactive to events and European policies (Weale *et al.* 2000; Liefferink and Jordan 2002), while domestic environmental politics is still a relatively closed, state-led process (Pridham and Kostadakopulos 1997; Botetzagias 2001). This mode of governance is sometimes called 'discretionary governance' (Hagendijk and Irwin 2006), where policy-making takes place with virtually no interaction with the general public or other major groups outside those institutions directly responsible for policy (essentially government departments and relevant industrial and scientific bodies). In this context, government is portrayed as serving universal goals of progress, welfare and growth. Previous research has identified this pattern in numerous science–environment debates in Greece, most notably in the case of biotechnologies (Marouda-Chatjoulis *et al.* 1998; Botetzagias *et al.* 2004; Reynolds *et al.* 2007).

Reflecting these trends, a small policy community consisting of relevant Ministries and energy producer groups (notably PPC) has tended to dominate Greek climate policy formulation. The First Greek Action Programme was elaborated under the responsibility of the Ministry for the Environment in collaboration with the Ministry for Development, while other competent Ministries, public sector bodies and private sector experts participated in developing the plan (SNC 1997). The SNC was prepared by nine ministries and two public bodies (PPC and the

Public Natural Gas Corporation) (SNC 1997), while co-ordination of the implementation of the EU emissions trading directive was given to a seven-person inter-ministerial committee (FONC 2006). The Office of Emissions Trading (within the Ministry for the Environment) was then created to monitor the implementation of the directive. Lastly, Greece's 2006 NAP was prepared in consultation with key sectoral bodies, and the Ministry for the Environment organized just one workshop for affected enterprises to allow an exchange of views on the NAP. Beyond these selective consultation processes, the exclusion of other societal groups was reinforced by the framing of climate policy as a techno-scientific issue rather than allowing open debate on measures to curb electricity demand.

On the other hand, little evidence exists of public willingness to support more ambitious emissions cuts. Environmental problems have always been a low politics issue in Greece compared with the economy and security. Public interest in environmental protection actually declined during the 1990s, while for the second half of the decade only one per cent of Greeks considered the 'Environment and the Quality of Life' to be important problems for Greece (Botetzagias 2001). Eurobarometer opinion polls did show that, between the mid-1980s and 1990s, Greeks were among the EU's most worried citizens about the greenhouse effect (81 per cent in 1996, EC average 69.9 per cent) (Eurobarometer 1997), yet, similar to other countries, this concern failed to materialize into concrete actions. In the same survey Greece recorded one of the highest percentages of individuals who had not taken a single personal action to save energy (33 per cent in 1996, EC average 24.9 per cent) (Eurobarometer 1997), while more recent research confirmed that Greeks were the least likely Europeans to have undertaken, or have the intention to undertake, energy-saving measures (Eurobarometer 2006). Coupled with energy considerations related to consumer purchases, Greeks also gained one of the lowest energy-saving index scores in the EU.

In response, political parties have tended to place little emphasis on environmental protection and climate change with the exception of the 2007 election, following catastrophic forest fires the preceding summer. Nevertheless, the ruling Conservatives' manifesto did not even mention climate change or renewable energy sources. Although the major opposition Socialist party's manifesto mentioned climate change as a significant challenge, it proposed mainly mainstream measures, for example saving energy, renewable energy and enforcing the polluter pays principle. Furthermore, responding to a Greenpeace questionnaire just before the 2007 election, the major Socialist and Communist opposition parties rejected (albeit for different reasons) the suggestion of a national target to reduce greenhouse gas emissions by 60–80 per cent below 1990 levels by 2050. The Conservatives, which won the elections, did not answer the questionnaire.

Interviews with NGO officials conducted for this research also confirmed the lack of discernible change in government on climate policy. Instead, most argued that climate change was considered to be too remote to influence any party's election strategy and that, instead, governments appeared to follow EU guidelines without showing any inclination to seize the political initiative. Another NGO interviewee contended that the state was applying 'an old kind of politics to a new kind of problem'. A representative from a major research institution commented further on the political costs of more aggressive action: 'Any government wishing to face this issue needs to be self-aware... that it will serve for a single term', while the stance of opposition party pledges on climate change were characterized by some interviewees as 'insincere', 'incidental', 'wishful thinking' and inversely proportional to the party's chances of forming a government.

Any other roads ahead?

The Greek energy *zugzwang*: prometheanism, energy demand and renewables

The domination of Greek greenhouse gas emissions by PPC-run lignite power generation plants combined with rising household energy demand and falling electricity prices means that any aggressive measures to curb emissions are likely to have significant implications for voters. Not surprisingly, therefore, successive governments have opted to prioritize supply-side energy 'rationalization' (natural gas and renewables) rather than tackling demand-side measures that may impose costs on final consumers or undermine PPC's interests (51 per cent of which still belongs to the state despite a series of privatizations). For example, in a Parliamentary debate on lignite's exemption from fuel taxes in late 2006, even the Socialist opposition party welcomed the government's retreat from a proposed lignite tax of €0.3 per KWh, noting that the duty (which was estimated to raise €110 million per year) would 'cause PPC's share to crash' and 'roll the extra cost over to the shoulders of the poor and miserable Greek people' (Hellenic Parliament 2006).

The Greek government could, thus, be said to be in a state of *zugzwang* in respect of promoting more ambitious climate policies. *Zugzwang* is a German chess term meaning 'compulsion to move' used to describe situations where any move by a player will result in the loss of a piece or a weakening of position. The first pin of the Greek climate policy *zugzwang* is the country's spiralling energy demand. The IEA (2002) projects that energy demand in Greece will rise by 4.5 per cent annually over the decade ending 2010, while the *First Report for the Long-term Energy Planning for Greece: 2008–2020* (Ministry for Development 2007) forecasts an increase in mainland electricity demand of between 38.8 and 55.7 per cent by 2020, depending on weather scenarios. The report also projects that solid fuel (mainly domestic lignite supplemented by imported bituminous coal) will contribute between 35 and 45 per cent of total electricity production under all main scenarios.

The second pin of the *zugzwang* is the near inevitability of political damage for any party that seeks to alter these dynamics. Hence, the most frequent strategy has been to substitute concrete action with rhetoric. For instance, when the Ministry for the Environment issued PPC with a €1 million fine for exceeding limits and failing to monitor sulphur dioxide and nitrous oxide emissions at some of its units – an action which made the front pages of all the Greek newspapers – Environment Minister Souflias made a largely spurious link between these fines for breaches of other atmospheric pollution regulations and climate change (Ministry for Environment 2007a). Shortly afterwards, the Regulatory Authority for Energy (RAE) sanctioned the construction of Greece's first two private bituminous coal plants.

To a large extent this has been a self-imposed *zugzwang* originating from a strong 'pro-supply' and econocentric bias in Greek energy management as well as a number of administrative constraints. For instance, local authority taxes, a real estate tax, and television-licensing fees are all collected through PPC electricity bills. Thus, a wholesale shift toward independent energy production would require the creation of a new tax-collection mechanism. The 'pro-supply' logic is also apparent in discourses surrounding renewable energies which, across environmental NGO, government and PPC publicity (Greenpeace Greece 2007a), stress the need to expand renewable energy provision without constraining energy demand. For instance, the Hellenic Wind Energy Association's (ELETAEN 2007) web-page features the motto 'Wind Energy; Clean, Inexhaustible, Renewable; Energy Forever'. The Greek Association of RES Electricity Producers' (HELLASRES 2007) website notes that renewables 'are practically inexhaustible energy sources and lessen the dependence on exhaustible, conventional energy sources'. This framing of renewable energy discourse may appear fairly generic but, in Greece, may feed assumptions that renewable energy can cope with any amount of demand. From a political perspective, such a Promethean discourse (Dryzek 1997) has the convenience of bypassing issues of energy demand by focusing on technological supply-side fixes.

However, this modernizing approach has not gone unchallenged. Most investors were interested in large-scale wind parks, mainly on the Aegean islands, with distribution of surplus electricity to the mainland. Such projects have often been opposed by local inhabitants and environmental NGOs. Nationwide environmental NGOs are more supportive of these plans, albeit to different degrees, while scientists and environmentalists have attacked what they perceive as local NIMBY campaigns, technophobia (for instance, claims that electromagnetic radiation from wind turbines causes cancer) and free-riding at the expense of mainland communities affected by lignite-fired power plants. This has led to calls for improved public education on renewable energies (interview with renewable energy expert, 2 October 2007). Nevertheless, cases also exist of communities opposing wind farms for broader reasons, including nature and landscape protection; alternative land uses and local development; the scale of projects vis-à-vis local needs; and equity in the distribution of the benefits and costs of renewable energy facilities. Others have denounced what they see as capital speculation for generously subsidized renewable energy projects.

Government subsidies for renewable energy developments have also not gone unchallenged. In August 2007, the influential Technical Chamber of Greece (TEE), representing the technical and mechanical Guilds of Greece, demanded the cessation and immediate re-design of the existing Plan for the Development of Photovoltaic Systems (TEE 2007), arguing that it posed unreasonable financial burdens on the state and end users, contained excessive goals, and had not resolved a number of planning, technical, institutional and siting issues. TEE recommended that less emphasis should be placed on large schemes compared with supporting 'clearly domestic production . . . by self-producers . . . primarily [those residing on] the islands'.

The Minister of the Environment responded forcefully to these criticisms during a presentation on the Special Zoning Plan for renewable energy:

> *It is a pressing need to reduce the energy produced by solid fuels and to increase as much as possible the production of green energy*. For this reason, any obstacles... resulting from local level planning will not be allowed and will be dealt with at the central level. Each one of us has to realise that it is not possible to demand more and more energy produced by burning solid fuels in other areas... to seek a cleaner environment... and, at the same time, to react to any attempt to produce environment-friendly energy.
>
> (Ministry for Environment 2007b: 10) (original emphasis)

This strong-arm rhetoric is likely to lead to legal disputes between protestors and the Ministry that may further delay and undermine the Greek renewable energy strategy and progress towards its emissions targets.

Greek energy restructuring

Given the numerous constraints on direct government intervention in renewable energy planning to achieve deeper emissions cuts in Greece, a more subtle political strategy to achieve similar ends may be the further liberalization and restructuring of the Greek energy market. The chief appeal of this is that it would increase the sensitivity of energy producers to price differentials between fuels and, via the market mechanism, accelerate the shift to lower-carbon electricity generation. However, although electricity liberalization has produced generally benign environmental outcomes and lower electricity prices in the UK (Steen and Vrolijk 2002), careful management is needed to counteract the risk of private sector short-termism leading to increased emissions (Ringel 2003; Vielle and Vigier 2007).

Moves to liberalize and privatize Greek energy production have also met with considerable opposition. In 2007, a PPC business plan explored the reduction of PPC's share of the Greek energy market from its current 90 per cent to as low as 28 per cent by 2020. The same report also projected that lignite's share of electricity production would reduce from 62 per cent to between seven per cent and 23 per cent. These developments were linked to concerns about the availability and quality of existing lignite beds, market liberalization, and increased operating costs arising from the need to acquire additional EU emissions trading permits if Greek energy restructuring did not take place. This news triggered a furious reaction from PPC's trade-union (GENOP-DEI), which denounced the plan as aiming to 'break down PPC; to give away power [production] to private capital' as well as 'to annul PPC's social mandate and increase the electricity bills;...to down-rate and spirit away our national fuel, lignite... [thus] abandoning the area [where lignite is mined and burnt] and driving its people into unemployment' (GENOP-DEI 2007a; 2007b). Although both arguments were devoid of any serious environmental content, this dispute delayed the endorsement of PPC's business plan by six months and again highlights the constraints on any strategy that prioritizes climate concerns over other economic or Promethean discourses on limitless energy demand.

Demand-side measures

As was noted earlier, the Greek government's scope to curb energy demand through consumer taxation is severely constrained by the public's unwillingness to pay higher electricity prices. The main strategy adopted by the Ministry for Development to promote energy conservation has been advertisements stressing the personal economic benefits of energy saving rather than environmental arguments. The side effects of other policies, as well as the restructuring of the Greek energy market, also have some potential to influence energy demand.

It is improbable that the current government – which holds only a slender parliamentary majority – would opt for a head-on confrontation on taxes. Thus, and for as long as PPC remains under state control, prices will continue to be constrained. Latest newspaper reports have nevertheless predicted further PPC privatizations and subsequent price increases of up to 25 per cent, while the RAE recently recommended that the Ministry for Development repeal low electricity rates for Greek industry. These developments may be regarded as the government testing the water for inescapable – if modest – increases in household energy prices.

In 2006, the Ministry for Development sought to galvanize action to curb energy demand by issuing a special circular requiring the replacement of energy-intensive lamps with more energy efficient bulbs in all public buildings. PPC has also accompanied small price increases in 2006 with incentives entitling consumers to a flat-rate five per cent discount in energy bills for reducing electricity consumption by a minimum of six per cent. Realigning this measure to reward customers according to the proportion of energy saved (with or without a cap) may provide a more sustained incentive; nevertheless, this initiative cost PPC €3.3 million in the first half of 2007 (0.14 per cent of PPC revenue from electricity sales) (PPC 2007), and may signal a greater willingness by

PPC to engage with energy conservation issues. Other low-impact, yet symbolic, policies could also help to pave the way for more ambitious policies. One example would be a phasing out of incandescent lamps, possibly starting with the Aegean islands, justified by their dependence on oil-produced energy.

Pressure from the EU and broadening domestic participation

Whatever the domestic context of Greek climate policy, Greece is likely to face continued pressure from the EU to adopt stronger emissions targets and measures (Skjärseth and Wettestad 2002). Although Greece secured a relatively undemanding reduction target to 2008–2012 under the EU burden-sharing agreement, pressure may be applied via European Court of Justice proceedings for non-compliance with Community law, where Greece already has a reputation for being one of the EU's worst environmental transgressors (Börzel 2003). The Commission recently initiated proceedings against Greece for non-implementation of Directive 2006/32/EC on energy end-use efficiency and energy services, which requires national strategies to achieve a nine per cent energy saving target by 2016, and Greece also received a reasoned opinion in 2006 for failing to transpose Directive 2002/91 on the energy performance of buildings, which could save Greek public sector buildings alone 425 $MtCO_2e$ per year (Greenpeace Greece 2007b).

Aside from legal measures, future developments in the EU emissions trading scheme – particularly on national emissions allocations, permit auctioning, and the use of Kyoto flexibility mechanisms – are likely to be another key influence on Greek climate policy. Although Greece received little criticism for its first two NAPs (Grubb *et al.* 2005; Betz and Sato 2006), the Commission may be less sympathetic in the future if Greece does not make demonstrable progress toward meeting its emissions targets. Similarly, the vast majority of EU emissions allowances were issued free of charge during the first two phases of the scheme. However, according to one World Wildlife Fund (WWF) study (2006), marketing 10 per cent of Greek permits during phase two would provide the government with around €615 million to fund other mitigation and abatement projects. Since affected companies would almost certainly pass on additional costs to consumers, this could generate further incentives for household energy conservation. Similarly, tighter restrictions on the purchase of Clean Development Mechanism and Joint Implementation emissions credits to meet national targets may create further incentives for domestic action.

Although the effects of EU policies do not constitute political strategies in their own right, scope exists for the Greek government to deflect criticism of stronger domestic measures by framing them as necessary responses to EU integration. As was noted earlier, Greece has performed as a classic EU environmental laggard state by obstructing legislative proposals, seeking country-specific derogations and failing to implement EU requirements (Börzel 2005). However, a subtle change in emphasis that stresses the political and economic importance of cooperation with the EU's climate regime may help the government to limit the political damage caused by more ambitious policies. It may also assist in marginalizing factions within the country that are opposed to any measure they perceive to compromise national economic interests, although it may also stir up greater anti-EU sentiments.

In the final analysis, however, the achievement of deeper greenhouse gas emissions cuts in Greece will depend on more active engagement by governments with business and civil society actors. A recent study by the Centre for Renewable Energy Sources (CRES 2006) concluded that virtually all energy efficiency measures in Greece are currently 'legislative-normative', with scarcely any cooperative and information-education-training measures, even in the domestic sector. This situation, although consistent with Greece's closed style of environmental decision-making, has perpetuated exclusionary and oppositional climate politics. Since further emissions reductions are likely to require unpopular measures, more inclusive and open dialogue with societal stakeholders on the needs, goals and measures of effective climate policy is needed to overcome current opposition. This will be no easy task, nor will it deliver major emissions cuts in the short term. Yet it is probably the only way of moving toward greater consensus on more ambitious measures.

References

Betz, R. and M. Sato (2006), 'Emissions trading: lessons learnt from the 1st phase of the EU ETS and prospects for the 2nd phase', *Climate Policy* 6, 351–59.

Börzel, T. A. (2003), *Environmental Leaders and Laggards in Europe: Why There is (not) a 'Southern Problem'*, Aldershot: Ashgate.

Börzel, T. A. (2005), 'Pace-setting, foot-dragging and fence-sitting: member state responses to Europeanization', in Jordan, A. (ed.), *Environmental Policy in the European Union: Actors, Institutions and Processes*, 2nd edition, London: Earthscan, pp. 162–80.

Botetzagias, I. (2001), *The Environmental Movement in Greece, 1973 to the Present*, unpublished PhD thesis, Department of Politics, Keele University.

Botetzagias, I., M. Boudourides and D. Kalamaras (2004), 'Biotechnology in Greece', *Science, Technology and Governance in Europe Project,* Discussion Paper 10.
Buchner, B., M. Catenacci and A. Sgobbi (2007), 'Governance and Environmental Policy Integration in Europe: What Can We Learn from the EU Emission Trading Scheme?' *SSRN eLibrary*.
Cass, L. (2005), 'Norm entrapment and preference change: the evolution of the European Union position on international emissions trading', *Global Environmental Politics* 5 (2), 38–60.
Costa, O. (2006), 'Spain as an actor in European and international climate policy: from a passive to an active laggard?' *South European Society and Politics* 11, 223–40.
CRES (Centre for Renewable Energy Sources) (2006), *Energy Efficiency Policies and Measures in Greece,* www.odyssee-indicators.org/Publication/PDF/nr_grc_2006.pdf [28 November 2007].
Damro, C. and P. Méndez (2003), 'Emissions trading at Kyoto: from EU resistance to Union innovation', *Environmental Politics* 12 (2), 71–94.
DESMIE (Hellenic Transmission System Operation) (2006), *Monthly Bulletin of Electricity Balance concerning the Interconnected System,* desmie.acn.gr/up/files/energy_200612.pdf [28 November 2007].
Dryzek, J. (1997), *The Politics of the Earth: Environmental Discourses,* New York: Oxford University Press.
ELETAEN (Hellenic Wind Energy Association) (2007), *Welcoming Page,* http://www.eletaen.gr/ [11 October 2007].
Eurobarometer (1991), *Special Eurobarometer No 057: Public Opinion in the European Union on Energy in 1991,* Brussels: European Commission.
Eurobarometer (1997), *Special Eurobarometer No. 104: European Opinion and Energy Matters 1997,* Brussels: European Commission.
Eurobarometer (2002), *Special Eurobarometer No. 169: Energy: Issues, Options and Technologies,* Brussels: European Commission.
Eurobarometer (2006), *Special Eurobarometer No. 258: Energy Issues,* Brussels: European Commission.
European Environmental Agency (1996), *Climate Change in Europe,* Copenhagen: European Environment Agency.
Eurostat (2007), *Statistics in Focus: Environment and Energy 80/2007,* Brussels: European Communities.
FAPCC (First Action Plan for Climate Change) (1995), *National Action Plan for the Abatement of* CO_2 *and other Greenhouse Gas Emissions,* Athens: Ministry for the Environment, Urban Planning and Public Works.
Fisher, D. (2003), 'Global and domestic actors within the global climate change regime: toward a theory of the global environmental system', *International Journal of Sociology and Social Policy* 23 (10), 5–30.
FNC (1995), *First National Communication to the UNFCCC,* Athens: Ministry for the Environment, Urban Planning and Public Works.
FONC (2006), *Fourth National Communication to the UNFCCC,* Athens: Ministry for the Environment, Urban Planning and Public Works.
GENOP-DEI (2007a), *Announcement: They are Demising PPC,* www.genop.gr/index.php?option=com_content&task=view&id=239&Itemid=1 [10 December 2007].

GENOP-DEI (2007b), *Press release: PPC's Demise will not be Allowed*, www.genop.gr/index.php?option=com_content&task=view&id=240&Itemid=1 [10 December 2007].

Greenpeace Greece (2007a), *The Solutions [to Climate Change]*, www.greenpeace.org/greece/campaigns/91303/91345 [20 October 2007].

Greenpeace Greece (2007b), *Incentives for Energy Conservation in Buildings: Greenpeace's Suggestions*, www.greenpeace.org/raw/content/greece/press/118523/efficiency-buildings.pdf [05 November 2007].

Grubb, M., C. Azar and U. Persson (2005), 'Allowance allocation in the European emissions trading system: a commentary', *Climate Policy* 5, 127–36.

Hagendijk, R. and A. Irwin (2006), 'Public deliberation and governance: engaging with science and technology in contemporary Europe', *Minerva* 44, 167–84.

HELLASRES (Greek Association of RES Electricity Producers) (2007), *RES Advantages*, www.hellasres.gr/Greek/giati-ape/giati-ape.htm [11 October 2007].

Hellenic Parliament (2006), *Minutes of the Discussion Concerning the Transposition into National Law of Directive 2005/19/EC*, Athens: Hellenic Parliament.

IEA (International Energy Agency) (1998), *Energy Policies of IEA Countries, Greece 1998 Review*, Paris: IEA.

IEA (2002), *Energy Policies of IEA Countries, Greece 2002 Review*, Paris: IEA.

IEA (2006), *Energy Policies of IEA Countries, Greece 2006 Review*, Paris: IEA.

Liefferink, D. and A. Jordan (2002), *The Europeanisation of National Environmental Policy: A Comparative Analysis*, Nijmegen: University of Nijmegen School of Management.

Marouda-Chatjoulis, A., P. Stathopoulou and G. Sakellaris (1998), 'Greece', in Durant, J., M. Bauer and G. Gaskell (eds), *Biotechnology in the Public Sphere: A European Sourcebook*, London: Science Museum, pp. 77–88.

Ministry for Development (2007), *First Report for the Long-Term Energy Planning for Greece: 2008–2020*, Athens: Ministry for Development.

Ministry for Environment (2007a), *Press release: A 1 Million Euro Fine to PPC for Atmospheric Pollution*, www.minenv.gr/download/2007-09-24.prostima.se.deh.doc [10 November 2007].

Ministry for Environment (2007b), *Press Conference by the Minister for the Environment, Mr. G. Souflias, on the Special Zoning Plan for the RES Siting*, www.minenv.gr/download/2007-02-01.g.souflias.parousiasi.idikoy.xorotaksikoy.sxedioy.gia.a.e.p.doc [10 November 2007].

Padilla, E. and J. Roca (2004), 'The proposals for a European tax on CO_2 and their implications for intercountry distribution', *Environmental and Resource Economics* 27, 273–95.

PPC (Public Power Corporation) (2007), *Press Release: Results for the First Semester of 2007*, www.dei.gr [28 July 2007].

Pridham, G. and D. Kostadakopulos (1997), 'Sustainable development in Mediterranean Europe? Interactions between European national and sub-national levels', in Baker, S., M. Kousis, D. Richardson and S. Young (eds), *The Politics of Sustainable Development: Theory, Policy and Practice within the European Union*, London: Routledge, pp. 127–51.

Rapanos, V. and M. Polemis (2005), 'Energy demand and environmental taxes: the case of Greece', *Energy Policy* 33, 1781–8.

Reynolds, L., B. Szerszynski, M. Kousis and Y. Volakakis (2007), 'GM-food: The role of participation in techno-scientific controversy', *Participatory Governance and Institutional Innovation [PAGANINI] Project Report*, Work Package 6.

Ringel, M. (2003), 'Liberalising European electricity markets: opportunities and risks for a sustainable power sector', *Renewable and Sustainable Energy Reviews*, 7, 485–99.

Ringius, L. (1999a), *The European Community and Climate Protection: What's Behind The Empty Rhetoric?* Oslo: Centre for International Climate and Environmental Research.

Ringius, L. (1999b), 'Differentiation, leaders, and fairness: negotiating climate commitments in the European Community', *International Negotiation* 4, 133–66.

SAPCC (Second Action Programme for Climate Change) (2002), *Second Action Plan for Curbing GHG Emissions (2000–2010)*, Athens: Ministry for the Environment, Urban Planning and Public Works.

Skjärseth, J. and J. Wettestad (2002), 'Understanding the effectiveness of EU environmental policy: how can regime analysis contribute?' *Environmental Politics* 11 (3), 99–120.

Skjärseth, J. and J. Wettestad (2007), 'Is EU enlargement bad for environmental policy? Confronting gloomy expectations with evidence', *International Environmental Agreements: Politics, Law and Economics* 17, 263–80.

SNC (1997), *Second National Communication to the UNFCCC*, Athens: Ministry for the Environment, Urban Planning and Public Works.

Steen, N. and C. Vrolijk (2002), 'United Kingdom: power markets and market policies', in Vrolijk, C. (ed.), *Climate Change and Power: Economic Instruments for European Electricity*, London: The Royal Institute of International Affairs, pp. 224–56.

TEE (Technical Chamber of Greece) (2007), *Press release: Suggestion for a Halt and an Immediate Re-Designing of the Plan for the Development of Photovoltaic Systems*, portal.tee.gr/portal/page/portal/PRESS/DELTIA_TYPOY/deltia2007/20070730 fotovolt.doc [30 August 2007].

TNC (2003), *Third National Communication to the UNFCCC*, Athens: Ministry for the Environment, Urban Planning and Public Works.

Vielle, M. and L. Viguier (2007), 'On the climate change effects of high oil prices', *Energy Policy* 35, 844–9.

Weale, A., G. Pridham, M. Cini, D. Konstadakopulos, M. Porter and B. Flynn (2000), *Environmental Governance in Europe: An Ever Closer Ecological Union?* London: Oxford University Press.

WWF (World Wildlife Fund) (2006), *European Economists and WWF ask for a Tougher ETS*, politics.wwf.gr/index.php?option=com_content&task=view&id=846&Itemid=381 [9 November 2007].

11
Facing up to the Greenhouse Challenge? Australian Climate Politics

Ian Bailey and Sam Maresh

Introduction

As a country, Australia is particularly vulnerable to the effects of human-induced climate change, particularly droughts and forest fires (CSIRO 2007). Despite this, Australians currently top the world league table of per capita greenhouse gas emitters. In 2004, annual per capita emissions in Australia stood at 27.2 tonnes of carbon dioxide equivalent, more than double the OECD country average (Turton 2004). Four main activities account for this outcome: (i) a high reliance on fossil fuels in the electricity generation mix; (ii) intensive road-based transportation within Australia's sprawling cities; (iii) its highly polluting non-ferrous metals sector, especially aluminium, which draws its energy mainly from subsidized and abundant coal reserves; and (iv) Australia's vast agricultural sector, which comprises 16 per cent of 2005 national emissions, far higher than in most other industrialized nations.

Australia was one of the earlier OECD countries to initiate a domestic climate strategy in 1992, but has since lagged behind most other affluent democracies in its commitments to address climate change (Papadakis and Grant 2003). This has been evident in a heavy reliance on voluntary programmes (Hamilton 2002; Taplin 2002) and former Prime Minister John Howard's opposition throughout most of his period in office to binding emissions targets without parallel commitments from developing nations (Grubb *et al.* 1999). However, recent developments suggest a re-engagement by Australian governments with climate issues, albeit still strongly mediated by national economic concerns. In 2007 and under strong pressure from the state governments, the Howard government announced the introduction of a national emissions trading scheme from 2011 (Australian Government 2007a; 2007b). Moreover, in November 2007 the new Labor administration, led by Kevin Rudd, honoured the first part of an election manifesto littered with commitments on climate policy by making ratification of the Kyoto Protocol its first official act in government.

To describe and explain these developments, the next section reviews the history and key features of Australian climate policy. This is followed by an analysis of the chief obstacles to more ambitious climate policy in Australia. Conclusions are then offered on political strategies that may enable Australian governments to make greater progress in curbing emissions.

Federal Australian climate policy 1992–2007

1992–2000: Early promise and false hopes

Christoff (2005) argues that Australia's initial response to climate change was largely informed by an altruistic concern for environmental protection and involved little serious contemplation of the economic implications of reducing Australian emissions. The government adopted a cooperative stance towards the United Nations Framework Convention on Climate Change (UNFCCC), and in 1992 released its first national climate strategy, the National Greenhouse Response Strategy (NGRS). This articulated a range of low- and no-cost emissions-reduction measures and an 'interim planning target' to stabilize emissions at 1988 levels by 2000. This was to be followed by a decrease in emissions of 20 per cent by 2005 (Commonwealth of Australia 1992).

To achieve these goals, the NGRS targeted improved understanding of climate-change impacts on Australia, national greenhouse gas accounting, and emissions reductions from all sources, sectors and sinks. This coincided with an ongoing public debate on carbon taxes; however, opposition to taxes from the influential resources sector on competitiveness grounds persuaded the Labor government led by Paul Keating that voluntary initiatives were a less politically damaging way to co-opt industry groups to take action (Taplin 2004). The Greenhouse Challenge (the Challenge) was accordingly introduced in 1995 as a joint initiative between the government and industry to promote voluntary, non-discriminatory and cost-effective action to cut emissions across all sectors of the economy (AIGN 2002).

Organizations that joined the Challenge signed partnership agreements to report on emissions and undertake action plans, though, crucially, members were only required to report against historic emissions, not to set or meet emissions targets (Taplin 2004). Although the government claimed that this approach would maximize participation and increase the aggregate impact of the Challenge, the NGRS was condemned as a weak response to climate change:

> Most programs [in the NGRS] were enhanced versions of pre-existing programs developed for reasons other than global warming... simply tack the greenhouse reduction objective onto existing programs... [and] have typified the Federal Government's approach to greenhouse policy through the 1990s.
>
> (Hamilton 1996: 4)

Following its election victory in 1996, the new National-Liberal Coalition led by John Howard replaced the NGRS with the National Greenhouse Strategy (NGS) in 1998, and created the Australian Greenhouse Office (AGO) as an executive agency of the Department of the Environment and Heritage to co-ordinate the NGS (Commonwealth of Australia 1998). In so doing Howard stressed the need for international and national climate policy to take into account Australia's strong reliance on fossil fuels and export-based economy, points reflected in the NGS's continued emphasis on voluntary measures and soft targets (Crowley 2007). The main components of the NGS are shown in Table 11.1.

Despite the introduction of new initiatives under the NGS, the Senate Environment, Communications, Information Technology and the Arts Committee (SECITAC) offered a damning assessment of the NGS and, in particular, the Greenhouse Challenge (SECITAC 2000). SECITAC agreed that the Challenge had stimulated some awareness-raising and practical efficiency measures, but concluded that it made no clear distinction between business-as-usual improvements in energy efficiency and extra efforts in response to government programmes, created no market disadvantages for non-participants and gave no incentives for companies to set or meet emissions targets (Parker 1999).

2001–2005: Stagnation and procrastination

The Australian government's stance on climate policy continued to harden throughout the late 1990s and early 2000s, and in 2002 Howard announced that Australia would not ratify the Kyoto Protocol because it exempted developing countries from binding emissions targets. This ushered in a period of relative stagnation in Australian climate policy, although Howard confirmed that Australia intended to honour its Kyoto target to restrict national emissions to 108 per cent of 1990

Table 11.1 Main elements of the National Greenhouse Strategy

Programme	Start date	Description	Emissions savings (MtCO2e) where specified
National Greenhouse Gas Inventory (NGGI)	1992	Annual estimates of greenhouse gas emissions based on the accounting rules applied to Australia's Kyoto Protocol emissions target	
Greenhouse Challenge	1995	Voluntary partnership programme between the government and industry to improve energy efficiency and reduce greenhouse gas emissions	21 $MtCO_2e$ per annum by 2002 against business-as-usual scenario
Greenhouse Challenge Plus	2005	As above	15 $MtCO_2e$ per annum by 2010 (Department of the Environment and Heritage 2003; 2005a)
Greenhouse Gas Abatement Programme (GGAP)	1998	Financial assistance to large-scale emissions-saving and sink-enhancing activities	6.1 $MtCO_2e$ by 2010 (Australian Government 2006a)
Generator Efficiency Standards (GES)	2000	Requires generators to monitor emissions and achieve technologically feasible and affordable industry best practice	
Mandatory Renewable Energy Target (MRET)	2001	Requires generators to take 2% or 9500 GWh of electricity from renewable sources by 2010	6.5 $MtCO_2e$ by 2008–2012 (Australian Government 2006a)

levels by 2008–2012 despite the decision not to ratify (AGO 2002). Further new initiatives were unveiled, including the Mandatory Renewable Energy Target (MRET), which requires generators to take two per cent or 9500 GWh of marketed electricity from renewable sources by 2010 (Australian Government 2003). However, the 9500 GWh cap, combined with increased energy demand, effectively reduced the MRET target to one per cent (Kent and Mercer 2006; MacGill and Outhred 2007).

The Challenge was also re-launched as Greenhouse Challenge Plus (Challenge Plus) in 2005. Challenge Plus largely replicates the Challenge's voluntary approach except that membership is now mandatory for all proponents of large-scale energy projects, organizations receiving over AUS$ three million per year in business fuel credits, and energy generators that meet the requirements of Generator Efficiency Standards (GES) (Department of the Environment and Heritage 2005a). Under Challenge Plus, GES participants are required to sign five-year binding agreements to assess their operations against best-practice guidelines for their technology classes and fuel types, though implementation of assessments is only required if projects meet government and industry 'economic justifiability' criteria. A new membership tier was also created to provide extra recognition for organizations that actually set and meet emissions targets (Sullivan 2005).

However, these measures failed to stem criticism of the Howard government's climate policies. Pollard (2003), for instance, estimates that only AUS$265 million of the AUS$ one billion expenditure planned under the Greenhouse Gas Abatement Programme (GGAP) was spent by 2003–2004, although the government calculated that GGAP projects would deliver 6.1 million tonnes of CO_2 equivalent ($MtCO_2e$) reductions by 2010 (Australian Government 2006a). GGAP was replaced in 2004 by a new Low Emissions Technology Demonstration Fund, which will operate until 2020 to support the commercial demonstration of technologies with the potential to deliver large-scale emissions reductions in the energy sector. This fund will receive AUS$500 million from government and aims to leverage a further AUS$ one billion from industry (Department of the Environment and Heritage 2005b).

A further sign of the stagnation in Australian climate policy under the Howard government was a protracted but largely sterile debate on emissions trading during the 1990s and early 2000s. The government commissioned two reports in 1999 and 2002, but both stressed the detrimental economic impacts of emissions trading on energy- and trade-dependent sectors without offering insights on how to overcome them (AGO 2002). The government then appeared to rule out emissions trading in its 2004 White Paper, *Securing Australia's Energy Future* (Australian Government 2004: 25):

> Australia will not impose significant new economy-wide costs, such as emissions trading, in its greenhouse response at this stage. Such action is premature in the absence of effective longer-term global action on climate change. Pursuing this path in advance of an effective global response would harm Australia's competitiveness and growth with no certain climate change benefits.

Frustration at John Howard's intransigence on climate policy combined with political motivations – prior to the 2007 election, all the state and territory governments were controlled by the Labor Party – led several state governments to develop independent climate initiatives to pressurize the federal government to take firmer action. Table 11.2 summarizes key state schemes currently in operation; however, their emergence has created marked regulatory overlap between state and federal policies, raising compatibility concerns if more concerted national action is taken. Another key state initiative was the setting up of the National Emissions Trading Taskforce by the state and territory governments in 2004 (State and Territory Governments 2004) to discuss a state-led emissions trading scheme in the absence of a clear policy lead from the federal government.

A final contentious aspect of the Howard administration's climate policies was its repeated claim that Australia was one of only a very few countries that were on course to meet their Kyoto Protocol targets (Commonwealth of Australia 2000; Australian Government 2005). During the Kyoto negotiations, Australia petitioned successfully for the insertion of Article 3.7, which enabled Annex B countries where land use and forestry are a major net source of emissions to include emissions from land-use changes in their 1990 emissions base year. The 'Australia clause', as it became known, effectively allowed Australia to inflate its base-year emissions knowing that land clearing had already been in decline throughout the 1990s, and that the clause effectively allowed other sectors to continue on a business-as-usual trajectory (Hamilton and Vellen 1999; Hunt 2004). Table 11.3 shows that, between 1990 and 2005, emissions from land-use change and forestry fell by 95.2 $MtCO_2e$ (–73.9 per cent). However, emissions from most other sectors rose sharply, by 104 $MtCO_2e$ (36.3 per cent) in the case of energy (AGO 2007; Crowley 2007).

Table 11.2 State-based mandatory emissions reduction schemes

Scheme	Start date	Target sectors	Description	Penalty AUS$[a]	Target emissions savings and dates
New South Wales Greenhouse Gas Abatement Scheme (NSW GGAS)	2003	Large electricity retailers and electricity customers	State benchmark of 7.27 tCO_2e per capita, 5 % below 1990. Sectors must meet intensity targets based on assigned proportion of total state electricity demand by surrender of Greenhouse Gas Abatement Certificates corresponding to attributed emissions.	12 per excess tCO_2e	26 $MtCO_2e$ savings achieved by 2007; extended to 2012–2020 subject to national emissions trading
Queensland Gas Scheme (QGS)	2005	Electricity retailers	Diversify state energy mix based on mandatory targets for purchase of electricity from eligible gas sources. Surrender of tradable gas electricity certificates within compliance periods.	11.40 per excess MWh	13 % of electricity generated from eligible gas, 18 % by 2020; 26 $MtCO_2e$ by 2020
Victorian Renewable Energy Targets (VRET)	2007	Electricity retailers, wholesale buyers	Mandatory and escalating targets for purchase of electricity from renewable sources by acquisition, surrender and trading of renewable energy certificates within compliance periods.	43 per excess MWh	10 % of electricity generated from renewable sources, saving 27 $MtCO_2e$ by 2030

[a] 2007, indexed to inflation.
Sources: MacGill *et al.* (2006); Queensland Government (2007); Victorian Essential Services Commission (2007).

Table 11.3 Australian greenhouse gas emissions 1990–2005 ($MtCO_2e$)

	Emissions $MtCO_2e$		**Change in emissions (%) 1990–2005**
	1990	**2005**	
Energy	287.0	391.0	36.3
Stationary energy	196.0	279.4	42.6
Transport	61.9	80.4	29.9
Fugitive emissions	29.1	31.2	7.3
Industrial processes	25.3	29.5	16.5
Agriculture	87.7	87.9	0.2
Land use, land-use change and forestry	128.9	33.7	–73.9
Waste	18.3	17.0	–6.9
Net emissions	**547.1**	**559.1**	**2.2**

Source: Australia's National Greenhouse Accounts.

From 2005: renewed engagement?

Following nearly a decade of reliance on 'no-regrets' measures and a defensive stance in international negotiations, the signs since 2005 are that Australia has begun to engage more robustly with the issue of climate change. This period began with renewed attempts by the Howard government to steer international climate politics towards the Australian national economic interest; however, domestic political pressure on Howard, combined with a landslide election victory in 2007 by the Labor Party under the leadership of Kevin Rudd, has generated new momentum for more ambitious climate policies within Australia.

On the international stage, the Howard administration embarked on a sustained campaign to influence global climate policy. In 2005, along with the USA, China, India, Japan and South Korea, Australia founded the Asia-Pacific Partnership on Clean Development and Climate (APP), an international non-treaty agreement to promote the development of technological solutions to climate change (APP 2006). Although the APP has been criticized for undermining the Kyoto Protocol and for relying on voluntary measures, its partners argue that it promotes practical cooperation between major developed and developing countries on climate change (Dennis 2006). To date, Australia has committed AUS$100 million to fund projects under eight international government and business taskforces established by the APP.

At the 2007 Asia-Pacific Economic Cooperation (APEC) meeting, Australia also secured the endorsement of the 21 APEC leaders (including the USA, China, India, Japan and Russia) for the *Sydney Declaration on Climate Change, Energy Security and Clean Development.* Although the declaration stopped short of specifying quantified emissions targets, it delivered a non-binding APEC-wide commitment to improve energy efficiency by at least 25 per cent by 2030 and increase forest cover by 20 million hectares by 2020, while also building strategic momentum for technological responses, flexibility and recognition of domestic circumstances faced by individual nations (APEC 2007). However, attempts by Australia to use APEC to establish emissions targets for developing nations were thwarted by China, which insisted that any targets must be negotiated through the UNFCCC (People's Republic of China 2007).

Finally, in 2007 the government established the Global Initiative on Forests and Climate. This initiative aimed to work with developing countries to support new plantations, limit deforestation and promote sustainable forest management (e.g. by improving fire-fighting capacity in Indonesia), and supported the Howard government's position on the importance of land-use and forestry changes in international climate policy (Australian Government 2007c). Australia has also highlighted the value of carbon credits associated with forestry to its regional neighbours (Turnbull 2007), although critics claim that illegal logging in some regions may make real emissions savings difficult to achieve.

By 2006, however, the Howard government was also coming under increasing pressure to reform domestic climate policy. Even ardent industrial supporters of voluntary programmes had begun to talk openly about emissions trading, and in December 2006 Howard established the Prime Ministerial Task Group on Emissions Trading made up of senior representatives from Australian business and public service to advise on the design of a national emissions trading scheme. In contrast with *Securing Australia's Energy Future,* the taskforce concluded that emissions trading was the most cost-effective option available to cut power and industrial emissions and that waiting for a new global deal on climate change might adversely delay or risk investment (Australian Government 2007a). Shortly afterwards, the government released *Australian Climate Change Policy,* which committed Australia to a national cap-and-trade scheme by 2011 (Australian Government 2007b). The key features of these proposals included:

- Maximum practical coverage of all sources, sinks and greenhouse gases (33 per cent of national emissions in 2010, rising to 45 per cent in 2015);

- A mix of free allocation (to protect trade-exposed and emissions-intensive sectors) and auctioning of single-year emissions permits;
- Initial exclusion of agriculture and land use;
- Capacity to link to other comparable national and regional schemes;
- Use of revenue from permits and fees to support low-emissions technologies and energy efficiency programmes.

However, the Howard government postponed any decision on a long-term 'aspirational' emissions goal, a crucial non-decision in terms of providing markets with an indication of the timing and pathways to reduced emissions.

Climate policy featured prominently in the general election campaigns of the two major parties for the first time in 2007. The Coalition first announced a new national Clean Energy Target (CET) scheme to generate 30,000 GWh each year from 'low emissions' power by 2020 (Howard 2007a), a definition that theoretically allows coal generation combined with carbon capture and storage to be classified as low emissions. CET is intended to replace all existing and proposed renewable energy schemes from January 2010; a move that should simplify the current patchwork of federal and state schemes but may also make some redundant well before CET comes into operation.

The Labor Party's manifesto also included a swathe of promises on climate policy, which the new Prime Minister Kevin Rudd will now face pressure to honour following his election victory in November 2007. Labor's term of office certainly began with a flurry of activity: ratification of the Kyoto Protocol; the creation of a new Department for Climate Change in December 2007; and the commissioning of the Garnaut Climate Change Review to examine the impacts, challenges and opportunities of climate change for Australia, the final report of which is scheduled for release in September 2008. Rudd also reaffirmed his party's commitment to emissions trading from 2010 and the use of revenues from permit auctioning to help low-income households cope with increased energy prices (Australian Labor Party 2007a). Slightly ironically, the previous government's procrastination on emissions trading caused some energy-intensive industries to delay expansion plans (Potter and Hughes 2007), producing a positive (if temporary) reduction in emissions growth in some sectors. Industry lobbying and managing political demands will, nevertheless, still feature strongly in decisions on emissions caps and on which industries are sufficiently trade exposed to receive free allocations.

In terms of targets and other initiatives, Rudd committed to cut Australia's greenhouse gas emissions by 60 per cent from 2000 levels by

2050 and to generate at least 20 per cent of electricity from renewable energy sources by 2020 (Australian Labor Party 2007a). This target is to be achieved by bringing MRET and state initiatives into an expanded national scheme. However, Labor also pledged to phase out this target from 2020 once national emissions trading was able to support investment in renewables. Another key Labor commitment is the introduction of a National Clean Coal Fund (NCCF) to develop and commercialize low-emission coal technologies (Australian Labor Party 2007b). The NCCF will operate to 2015 and is accompanied by further support for the coal industry through the setting of national targets for electricity generation from clean coal technologies to encourage their early commercialization.

The nuclear option

Nuclear power has been a recent and controversial element of the climate policy debate in Australia following the commissioning of a taskforce in 2006 to consider the contribution of uranium mining, processing and nuclear energy to greenhouse gas reduction (Australian Government 2006b). Although Australia is a significant uranium exporter and holds 40 per cent of the world's known low-cost reserves, several state and federal laws prohibit nuclear power. The taskforce concluded that the use of nuclear power could deliver national greenhouse gas emission reductions of 8–17 per cent by 2050. However, concerns about the development and operating costs of nuclear power (even with emissions trading), the lack of a nuclear skilled workforce, existing legal prohibitions and a non-existent regulatory framework make it likely to be at least 15 years before nuclear electricity could be delivered to the Australian grid, although the report noted that Australia had sufficient suitable sites to accommodate the storage of nuclear waste.

John Howard argued that nuclear power was the only reliable source of low-emissions baseload power available to Australia and many other countries (Howard 2007b). However, the Labor Party consistently opposed nuclear power because of concerns about waste storage and its high demand on water resources (Australian Labor Party 2007c). Public and political opposition thus make it unlikely that nuclear power will contribute to Australia's energy-generation mix in the foreseeable future.

In summary, Australian climate policy has undergone three distinct phases since 1992: the early adoption of 'no-regrets' measures; a period of virtual stagnation during which economic arguments largely subsumed other considerations; and a more recent, though still nascent, resurgence of interest in market-based instruments and international cooperation. Throughout this evolution, however, the overwhelming political priority has been domestic economic interests, particularly those of the important resources sector. Tensions within Australia's federal system have also created a complex patchwork of state and federal requirements which, although innovative, has undermined the coherence of Australian climate policy. The next section analyses in more detail the main factors that have contributed towards the current situation.

Obstacles to more ambitious Australian climate policy

Fossil fuel dependency and economic self interest

With a few exceptions, the development of more ambitious climate mitigation policies in affluent countries is constrained to some degree by a historical dependence on fossil fuels for energy production and the perceived adjustment costs of moving to a low-carbon economy. However, Australia's rich endowment of lignite, black coal, natural gas and other mineral resources (notably iron ore and bauxite) has provided a cheap supply of raw materials for its energy-intensive steel and aluminium industries, making the Australian economy more reliant on fossil fuels than is the case in most advanced nations (Jessup and Mercer 2001; Geoscience Australia 2005). While scarcity in other natural resources (especially water) and concerns about salinization in the Murray–Darling Basin and Western Australia have amplified the political salience of climate change, until recently large-scale land clearance for agriculture was condoned in several states as a means of feeding Australia's growing population and generating export revenue.

Allied to this is Australia's geographical proximity to, and strong trade links with, the major Asian economies. During the 1950s and 1960s this took the form of government subsidization of 'import substitution' industries in sectors like plastics and chemicals aimed at restricting Australia's economic dependence on its politically unstable or 'unsavoury' East Asian neighbours (Hamilton and Turton 1999). This was coupled with the active promotion of export industries supplying unrefined or part-refined minerals and fossil fuels to Asia; firstly, Japan followed by the Asian 'tiger' economies and, more recently, China and India, many of which had few commercial incentives to reduce greenhouse gas emissions (Lyster 2004).

This combination of factors enabled the resources sector to form a powerful lobby that has repeatedly and successfully utilized warnings of an industrial exodus from Australia to press for cautious climate policies. Although some commentators claim that the country's abundance of cheap energy and minerals make it unlikely that industries such as aluminium will leave Australia (Turton 2002), their domination by trans-national corporations has enhanced the political credibility of such threats. As a result, Crowley (2007: 123) argues that a blurring of government-industry boundaries exists in 'the drafting and managing of policy, the funding of research into the cost of abatement, and the inclusion of industry in official international delegations'. Although the fossil fuel lobby's influence aligned conveniently with the Howard government's conservative politics and its coalition partner's resources sector interests, industry lobbyists also acted swiftly to persuade Paul Keating's Labor administration to abandon carbon taxes in favour of voluntary measures. The sector's blocking manoeuvres have also enjoyed considerable success in some resource-dependent states, particularly Western Australia.

Economic self interest has also largely dictated Australia's stance towards international climate policy. On the one hand, John Howard's refusal to ratify Kyoto unless developing countries also adopt binding targets will have antagonized those nations that stand to bear the brunt of climate-change impacts (Hunt 2004). On the other hand, the Coalition initiated a targeted diplomatic campaign with its regional trading partners to secure Australia's share of booming international resource markets, and used plurilateral fora like the APP and APEC to garner support among major developing and developed nations for global climate agreements that are effectively a replica of Australia's domestic policies: that is, broad in coverage but with strong concessions to national constraints; flexible rather than mandatory; and technologically driven. Australia's Kyoto status has now changed but how far its progress towards its Kyoto target will be affected by the policy foundations set out by the Rudd government remains to be seen.

The complexities of federal governance

A second explanation for the nature of Australian climate politics can be found in the country's federal system of governance and the resultant division of responsibilities for economic and environmental policy between the federal and state governments which this creates. Although the federal government has no explicit environmental powers, the

Australian High Court has conferred powers on Canberra in several state-federal environmental disputes. The Commonwealth also controls many key aspects of economic policy, particularly funding to the state governments (Crowley 2007). Federalism is often seen as means of allowing state governments to experiment with policies which, if successful, can be taken up as federal measures (Sbragia 1996). However, within Australia it has also led to a complex and fragmented patchwork of climate regulation that has helped to legitimate industry arguments about climate policy being prohibitively expensive (AIGN 2002). Additionally, climate politics has often been polarized along party lines, with state (Labor) governments supporting Kyoto ratification and using climate policy to pressurize, and assert autonomy from, the Howard administration (Fenna 2007). The outcome of this, as often as not, has been a lack of policy cohesion and long-term direction. Howard also proved extremely adept at using 'wedge tactics' to neutralize opponents while appealing to narrowly defined (usually economic and national security) conceptions of 'the national interest' (Crowley 2007). That said, state-federal tensions have led to policy innovation, such as the New South Wales Greenhouse Gas Abatement Scheme, while the Prime Ministerial Task Group on Emissions Trading was at least in part commissioned to recapture the political initiative following the setting up of the state-led National Emissions Trading Taskforce.

Electoral interests

Crowley (2007) reports the findings of several public opinion polls which suggest that climate change was an issue of concern to 78 per cent of Australians in 2003, with 80 per cent believing in 2001 that Australia should ratify the Kyoto Protocol regardless of the US position. Despite this, electoral support for more progressive Australian climate policy is not helped by the problems that a carbon-constrained economy pose for an electorate that has become accustomed to carbon-lavish lifestyles. Leigh (2005) also reports that economic, social and taxation issues continue to influence public voting more strongly than environmental issues. Thus, historically, the main political parties never felt obliged to make climate policy a priority election issue. The Climate Institute reported that 62 per cent of voters in marginal seats said that climate issues – made palpable by Australia's 'thousand year' drought and resulting crop failures, water shortages and bush fires – would influence their vote in the 2007 election (Crowley 2007). Nevertheless, a soft approach to climate policy has generally been sufficient to assuage electorates' climate concerns without compelling a significant deviation from the stance of the fossil fuel lobby. Similarly, while Labor skilfully reflected public sentiments about climate change in the 2007 election, whether electorates will accept measures that impact on their material freedoms remains less clear.

Of these obstacles, the most significant are undoubtedly the energy- and trade-dependent character of the Australian economy and the influence of industry groups on national and state climate politics. This does not mean that industry is uniformly ambivalent towards stronger measures or that government is entirely captured by special interests. However, what is clear is that the distinctive characteristics of the Australian economy prevent a simple reading across of European solutions. Even though Australia's siege mentality towards climate policy now appears to have lifted, these structural constraints are unlikely to dissipate, whichever political party holds office. With this in mind, the final section considers political strategies that may assist the introduction of more progressive climate policies.

Future strategies for Australian climate policy

Reframing the economic debate

A core priority for any Australian government seeking to persuade industry groups and the electorate to make a more active contribution to reducing greenhouse gas emissions is to challenge entrenched assumptions about the costs and benefits of climate policy. Federal administrations have historically framed climate policy as an economy-versus-environment debate and stressed the need for Australia to defend its industries against uneven competition from the non-carbon-regulated Asian mega-economies. Recent developments with emissions trading, where for the first time an economic instrument was adjudged to be more economically efficient than voluntary measures, suggests that political discourse is already changing. However, active government intervention is needed (utilizing the Stern Review and Intergovernmental Panel on Climate Change reports) to embed this 'win–win' discourse; much will also depend on the performance of the EU emissions trading scheme and lesson drawing from Europe's experiences with this instrument.

Another way in which such reframing might take place is through emphasizing Australia's long-term economic vulnerability if other countries decarbonize substantially and Australia remains reliant on diminishing fossil-fuel exports, or if a global carbon price is established and

Australia fails to reduce the carbon intensity of its aluminium and steel industries. Although Australia might adapt to these scenarios by consolidating its uranium exports, this carries substantial economic and political risks if influential countries reject nuclear energy or widespread public opposition to new uranium mines materializes. Taking a longer-term perspective that stresses the economic (as well as environmental) benefits of climate policy thus provides the first way in which the government can become disentangled from the economy-versus-environment debate that has tended to dominate Australian climate politics.

Hard carrots and soft sticks

Although reframing perceptions is an important and necessary first step in winning support for stronger climate policies, tangible incentives are also needed to stimulate energy efficiency and low-carbon energy sources where Australia has a natural advantage, such as photovoltaics. A pragmatic approach to this issue is through the development of 'strong carrots' and 'soft sticks' policies, where flexible and (now) accepted instruments like emissions trading (soft sticks) are accompanied by further expansion of funding programmes like the Low Emissions Technology Demonstration Fund (hard carrots) to commercialize low-emission technologies and soften industry resistance to mandatory targets for low-emissions energy sources. Similarly, industry may be more receptive to below business-as-usual emissions caps if a proportion of costs can be passed on to consumers (which would also help to raise public awareness of domestic energy consumption) and if trade-exposed industries are suitably compensated through capacity-building funding. Whatever the mix of trade-offs adopted, it must recognize the need to soften the significant restructuring that the Australian economy and society must undergo to achieve a low-carbon state. 'Soft stick' and 'hard carrot' policies appear to be an appropriate way of building momentum towards this goal.

Creating direction

A related political tactic to promote action is the creation of long-term emissions-reduction goals to reassure industry that investments in emissions reduction will produce financial rewards. An important first step in this direction has been made with the Rudd government's announcements of an overall target of a 60 per cent reduction in greenhouse gas emissions from 2000 levels by 2050, plus renewable energies targets. Another approach would be to set limits on significant new emissions sources such as baseload coal-fired power stations, a move that would be popular electorally and may further drive demand for low-emissions and renewable energies. The government must also decide how land-use changes will be accounted for in the future. Past governments have sought to maximize notional emissions reductions arising from avoided deforestation to offset the lack of progress by energy-intensive industries. An alternative tactic, similar to that applied to industry and power generation, would be to alter the costs and benefits to landowners of continuing agriculture on cleared marginal land compared with sink-enhancing activities through the development of domestic investment programmes and international initiatives like the Global Initiative on Forests and Climate. Nevertheless, any over-reliance on contested carbon-sink policies is likely to face stern public scrutiny.

Building political and federal consensus

A final priority for the Australian government is to address the cross-party divisions on climate policy – both in federal–state interactions and nationally – that have contributed towards a fragmented Australian climate policy and a more general failure to challenge industry preferences for voluntary measures. Such a political consensus is now theoretically closer since the election of the new federal Labor administration, since, at the time of writing, all the states are Labor controlled. However, the nature of Australian politics is such that state elections tend to return whichever political party is in opposition at national level. Labor's current dominance over state and federal politics may, thus, be short-lived.

One option to resolve this problem would be greater policy centralization. However, this would almost certainly be construed by state governments as an assault on their autonomy and as requiring an unjustified level of trust that the federal government will live up to its promises on climate policy and other issues. The current two-tier system has also stimulated considerable policy innovation and been a useful vehicle for creating pressure for stronger federal measures. Maintaining these creative 'checks-and-balances' therefore seems to be the more promising and politically acceptable strategy; however, a minimum pre-requisite must be the nurturing of a greater cross-party détente which recognizes that conservative domestic policies and an obstructive approach to international agreements is both economically *and* environmentally counter-productive to Australia's long-term interests.

Political strategizing notwithstanding, the Australian economy's structural dependency on carbon-based fuels – coupled with its strong trade links to the USA and Asian economies that have yet to commit to binding carbon constraints – present genuine difficulties for any Australian government seeking to make substantive cuts in greenhouse gas emissions. Alongside domestic initiatives, therefore, Australian climate politics will continue to be shaped strongly by its interactions with its APEC and APP partners, and in this regard Australian climate policy remains intimately tied to the stance of these countries and, in particular, to that of the USA.

References

AGO (Australian Greenhouse Office) (2002), *Pathways and Policies for the Development of a National Emissions Trading System for Australia*, Canberra: AGO.

AGO (2007), *National Greenhouse Gas Inventory 2005*, Canberra: AGO.

AIGN (Australian Industry Greenhouse Network) (2002), *No Disadvantage, No Discrimination*, Canberra: AIGN.

APEC (Asia-Pacific Economic Cooperation) (2007), *Sydney APEC Leaders' Declaration on Climate Change, Energy Security and Clean Development*, http://www.apecsec.org.sg/apec/leaders__declarations/2007/aelm_climatechange. html [13 November 2007].

APP (Asia-Pacific Partnership) (2006), *Charter for the Asia-Pacific Partnership on Clean Development and Climate*, Sydney: APP.

Australian Government (2003), *Renewable Opportunities: Review of the Operation of the Renewable Energy (Electricity) Act 2000*, Canberra: AGO.

Australian Government (2004), *Securing Australia's Energy Future*, Canberra: Department of the Prime Minister and Cabinet.

Australian Government (2005), *Australia's Fourth National Communication on Climate Change: Report under the UNFCCC*, Canberra: AGO.

Australian Government (2006a), *Greenhouse Gas Abatement Programme (GGAP)*, Canberra: Commonwealth of Australia.

Australian Government (2006b), *Uranium Mining, Processing and Nuclear Energy: Opportunities for Australia? Report to the Prime Minister by the Uranium Mining, Processing and Nuclear Energy Review Taskforce*, Canberra: Department of the Prime Minister.

Australian Government (2007a), *Report of the Task Group on Emissions Trading*, Canberra: Commonwealth of Australia.

Australian Government (2007b), *Australian Climate Change Policy: Our Economy, Our Environment, Our Future*, Canberra: Commonwealth of Australia.

Australian Government (2007c), *Global Initiative on Forests and Climate*, Canberra: AGO.

Australian Labor Party (2007a), *Election 07 Policy Document – Labor's 2020 Target for a Renewable Energy Future*, Canberra: Australian Labor Party.

Australian Labor Party (2007b), *New Directions for Australia's Coal Industry*, Canberra: Australian Labor Party.

Australian Labor Party (2007c), *Nuclear Power? No Thanks*, http://www.kevin07.com.au/fresh-ideas/climate-change-water/nuclear-power.html [8 November 2007].

Christoff, P. (2005), 'Policy autism or double-edged dismissiveness? Australia's climate policy under the Howard government', *Global Change, Peace and Security* 17, 29–44.

Commonwealth of Australia (1992), *National Greenhouse Response Strategy*, Canberra: Commonwealth of Australia.

Commonwealth of Australia (1998), *The National Greenhouse Strategy: Strategic Framework for Advancing Australia's Greenhouse Response*, Canberra: AGO.

Commonwealth of Australia (2000), *National Greenhouse Strategy: 2000 Progress Report*, Canberra: Commonwealth of Australia.

Crowley, K. (2007), 'Is Australia faking it? The Kyoto Protocol and the greenhouse policy challenge', *Global Environmental Politics* 7, 118–39.

CSIRO (Commonwealth Scientific and Industrial Research Organization) (2007), *Climate Change in Australia: Technical Report 2007*, Canberra: CSIRO.

Dennis, C. (2006), 'Promises to clean up industry fail to convince', *Nature* 439, 253.

Department of the Environment and Heritage (2003), *National Greenhouse Gas Inventory 2003*, Canberra: Department of the Environment and Heritage.

Department of the Environment and Heritage (2005a), *Greenhouse Challenge Plus: Programme Framework 2005*, Canberra: AGO.

Department of the Environment and Heritage (2005b), *Department of the Environment and Heritage Annual Report 2004–05*, Canberra: Department of the Environment and Heritage.

Fenna, A. (2007), 'The malaise of federalism: comparative reflections on commonwealth-state relations', *Australian Journal of Public Administration* 66, 298–306.

Geoscience Australia (2005), *Australia's Identified Mineral Resources*, Canberra: Australian Government/Geoscience Australia.

Grubb, M., C. Vrolijk and D. Brack (1999), *The Kyoto Protocol: A Guide and Assessment*, London: The Royal Institute of International Affairs.

Hamilton, C. (1996), *Why Government Won't Take Greenhouse Seriously*, Canberra: The Australia Institute.

Hamilton, C. (2002), *Running from the Storm: The Development of Climate Change Policy in Australia*, Sydney: University of New South Wales Press.

Hamilton, C. and H. Turton (1999), 'Subsidies to the aluminium industry and climate change', *Background Paper No. 21*, Canberra: The Australia Institute.

Hamilton, C. and L. Vellen (1999), 'Land-use change in Australia and the Kyoto Protocol', *Environmental Science and Policy* 2, 145–52.

Howard, J. (2007a), *National Clean Energy Target*, http://www.pm.gov.au/media/Release/2007/Media_Release24577.cfm [7 November 2007].

Howard, J. (2007b), *Address to the Liberal Party Federal Council 3 June 2007*, http://www.pm.gov.au/media/speech/2007/Speech24350.cfm [22 October 2007].

Hunt, C. (2004), 'Australia's greenhouse policy', *Australian Journal of Environmental Management* 11, 156–64.

Jessup, B. and D. Mercer (2001), 'Energy policy in Australia: a comparison of environmental considerations in New South Wales and Victoria', *Australian Geographer* 32, 7–28.

Kent, A. and D. Mercer (2006), 'Australia's mandatory renewable energy target (MRET): an assessment', *Energy Policy* 34, 1046–62.

Leigh, A. (2005), 'Economic voting and electoral behaviour: how do individual, local and national factors affect the partisan choice?' *Discussion Paper 489*, Centre for Economic Policy Research, Canberra: Australian National University.

Lyster, R. (2004), 'Common but differentiated? Australia's response to climate change', *Georgetown International Environmental Law Review* 16, 561–91.

MacGill, I. and H. Outhred (2007), 'Australian climate policy and its implications for AP6 countries', *China Energy Law International Symposium*, Beijing.

MacGill, I., H. Outhred and K. Nolles (2006), 'Some design lessons from market-based greenhouse gas regulation in the restructured Australian electricity industry', *Energy Policy* 34, 11–25.

Papadakis, E. and R. Grant (2003), 'The politics of "light-handed regulation": "new" environmental policy instruments in Australia', in Jordan, A., R. Wurzel and A. Zito (eds), *'New' Instruments of Environmental Governance: National Experiences and Prospects*, London: Frank Cass, pp. 27–50.

Parker, C. (1999), 'The Greenhouse Challenge: trivial pursuit?' *Environmental and Planning Law Journal* 16, 63–74.

People's Republic of China (2007), 'China and Australia issue joint statement on climate change', Chinese Embassy, Canberra, 7 September.

Pollard, P. (2003), 'Missing the target: an analysis of Australian government greenhouse spending', *Australia Institute Discussion Paper 51*, Canberra: The Australia Institute.

Potter, B., and D. Hughes (2007), 'Export boost smelting away', *Australian Financial Review*, 23 November, Sydney, 25.

Queensland Government (2007), *ClimateSmart 2050; Queensland Climate Change Strategy 2007: A Low-Carbon Future*, Brisbane: Queensland Government.

Sbragia, A. (1996), 'Environmental policy: the push-pull of policy-making', in Wallace, H. and W. Wallace (eds), *Policy-making in the European Union*, Oxford: Oxford University Press, pp. 235–55.

SECITAC (Senate Environment, Communications, Information Technology and the Arts Committee) (2000), *The Heat Is On: Australia's Greenhouse Future*, Canberra: Parliament of Australia Senate.

State and Territory Governments (2004), *Inter-Jurisdictional Working Group on Emissions Trading*, http://www.emissionstrading.nsw.gov.au/__data/assets/pdf_file/0005/410/terms.pdf [8 November 2007].

Sullivan, R. (2005), 'Code integration: alignment or conflict?' *Journal of Business Ethics* 59, 9–25.

Taplin, R. (2002), *Independent Verification of the Greenhouse Challenge Program: Final Report*, Sydney: Snowy Mountains Electricity Corporation, for Australian Greenhouse Office.

Taplin, R. (2004), 'Australian experience with greenhouse NEPIs', in Jacob, K., M. Binder and A. Wieczorek (eds.), *Governance for Industrial Transformation: Proceedings of the 2003 Berlin Conference on the Human Dimensions of Global Environmental Change*, Berlin: Environmental Policy Research Centre, pp. 491–501.

Turnbull, M. (2007), *High-Level Meeting on Forests and Climate Change*, Sydney: Australian Greenhouse Office.

Turton, H. (2002), 'The aluminium smelting industry: structure, market power, subsidies and greenhouse gas emissions', *Australia Institute Discussion Paper 44*, Canberra: The Australia Institute.

Turton, H. (2004), 'Greenhouse gas emissions in industrialised countries: where does Australia stand?' *Australia Institute Discussion Paper 66*, Canberra: The Australia Institute.

Victorian Essential Services Commission (2007), *Victorian Renewable Energy Target (VRET) Scheme*, http://www.esc.vic.gov.au/public/VRET [8 November 2007].

12
Explaining the Failure of Canadian Climate Policy

Douglas Macdonald

Introduction

In 2004, Canadian greenhouse gas emissions were approximately 27 per cent above those in 1990, well adrift of its Kyoto goal of reducing emissions to 6 per cent below 1990 levels by 2008–2012 (Canada 2007: 9). It is conceivable that a combination of draconian domestic policy and a massive purchase of international credits might allow Canada to reach its Kyoto goal. That objective, however, has been ruled out by the Canadian government. Since it was elected in January 2006, the Conservative Party government led by Prime Minister Stephen Harper has consistently said that Canada cannot – and indeed should not, given the economic cost – meet its Kyoto goal. Instead, in April 2007 his government introduced policy measures intended to achieve a 20 per cent reduction below the *2006 level* by 2020 (Environment Canada 2007). Since 2006 emissions were around 30 per cent above the 1990 level, the current Canadian objective is to reduce emissions to two per cent above the 1990 level (Pembina Institute 2007). Thus, Canada is officially trying to meet neither its Kyoto goal nor the 2012 deadline. Economic modelling by Rivers and Jaccard predicts that the Conservative government policy will not even meet this goal and that, instead, emissions will continue to rise until 2020 (Simpson *et al.* 2007: 196).

To explain this policy failure, this chapter first provides an account of the evolution of Canadian climate policy since it ratified the United Nations Framework Convention on Climate Change (UNFCCC) in 1992. At appropriate points in that history, I identify policy options which have been advocated but not adopted by federal and provincial governments; the periodic influence of international events upon domestic policy; and efforts by governments at both levels to take action and at the same minimize associated political costs. I then discuss briefly the four major explanations offered to date by Canadian analysts to explain the ineffectiveness of those policy actions. This is followed by my own analysis of what I see as the major explanatory factors: the lack of a provincial champion and the lack of federal government leadership, both of which stem ultimately from the weakness of Canadian environmentalism. I conclude by arguing that two things are needed to move Canadian policy onto a more effective track. The first is pressure by both environmentalists and external, international actors to keep the issue salient and blunt the lobbying power of business. The second is recognition by the rest of the country that it has to negotiate with the Saudi Arabia of Canada, the province of Alberta, and offer some form of compensation for the fact that effective Canadian policy will impose a higher price on that province than any other.

Canadian climate policy, 1992–2007

The Canadian constitution gives ownership of resources to the provinces, rather than the federal government. Although the courts have decreed that both levels of government hold jurisdiction on environmental issues, historically environmental regulation has been done almost exclusively by the provinces, with the federal government focusing on science and efforts to coordinate provincial policy. National climate policy thus requires coordinated action by both the federal and the provincial governments. During the period 1992–2002 that was done, albeit at the price of lowest-common denominator decision-making. With the federal government decision to ratify the Kyoto Protocol in 2002, however, the federal-provincial process ended and has not resumed. Since 2002, the federal and provincial governments have developed their climate policies unilaterally, with no attempt at coordination. The history which follows, accordingly, is divided into three parts: (i) the national process up to 2002; (ii) the battle over ratification in 2002 and (iii) unilateral federal government policy-making since then, with some reference to provincial policy. Table 12.1 provides a chronological overview.

When it hosted the Toronto conference in 1988, Canada helped to put the climate issue on the global policy agenda. Four years later, it pressed for adoption of the UNFCCC at Rio and, in December 1992, was one of the first countries to ratify the framework convention. In 1993, the Liberal government led by Jean Chrétien was elected on a strong environmental platform. Willingness to implement that platform was

Table 12.1 Key dates in Canadian climate policy

1988	Canada co-hosts Toronto conference
1990	Canadian government announces unilateral stabilization objective
1992	Canada ratifies UNFCCC
1995	Federal-provincial National Action Program on Climate Change
1997	Canada commits to reducing to 6 per cent below 1990 levels
2002	Canadian government decides to ratify
	Federal-provincial process ends
	First unilateral federal government plan
2005	Second unilateral federal government plan
2006	Federal government explicitly abandons the Kyoto goal
2007	Third federal government unilateral plan

soon blunted, however, by resistance from the oil and gas and other industries which would be adversely affected by rising energy prices or regulatory requirements to reduce greenhouse gas emissions, and by the fossil-fuel producing provinces, led by Alberta.

In 1993 and 1994, national policy development was coordinated by two intergovernmental secretariats, the Canadian Council of Ministers of the Environment (CCME) and the Council of Energy Ministers (CEM). Federal and provincial energy and environment ministers met on a regular basis to oversee a process of multistakeholder consultation. Within the federal government, there was a split between the Environment Minister, pressing for the use of more coercive policy instruments, and the Natural Resources Minister, who was also the Chrétien government's Alberta lieutenant and worked to protect the associated interests of that province and the oil industry. The latter won that battle, and since there were no provinces pressing for vigorous action, the 1995 Canadian national programme relied primarily upon the Voluntary Challenge and Registry Program (VCR) to bring about emission reductions by large industrial sectors, which account for approximately half of total emissions. At the time, Quebec's separatist Parti Québécois government was preparing for its second referendum on sovereignty-association and was not participating in any national programmes. Accordingly, Quebec acted independently to put in place its own comparable programme, *ÉcoGESte*.

During the 1993–1994 consultations, environmentalists advocated coercive policy instruments, both economic and regulatory, while business advocated voluntarism. Even before the consultations were completed, Prime Minister Chrétien ruled out a carbon tax in order to reassure a nervous Alberta (Simpson *et al.* 2007: 73), saying to a Calgary audience: 'Relax, relax... It's not on the table, and it will not be on the table' (Corcoran 1994). During the mid-1990s his government and the provinces were significantly reducing spending in order to constrain rising annual deficits and accumulated debt, thus reducing the policy capacity of environment and resource departments. This, coupled with Alberta resistance and the relatively low public salience of the issue, explains governments' adoption of voluntarism as the principal policy instrument. The political strategy adopted was to give the appearance of action, while avoiding alienating powerful political actors.

In 1997, as Canada and other countries prepared for the third UNFCCC Conference of the Parties (COP), to be held in Kyoto, the machinery of federal-provincial national policy-making generated an agreement on the position to be taken into the Kyoto negotiations – Canada should keep the goal of emissions stabilization at the1990 level, but extend the deadline to 2010. Immediately prior to the Kyoto meeting, however, Prime Minister Chrétien personally intervened in the climate policy process, responding to pressure from other heads of state, most notably US President Bill Clinton (Harrison 2006). As a result, Canada accepted a more ambitious objective at Kyoto, to reduce emissions to 6 per cent below 1990 levels by 2008–2012. Not surprisingly, this provoked a strong protest from Alberta, which in turn led the federal and provincial governments to adopt the principle that national climate policy would be designed in a way that did not impose an unacceptable cost upon any one region of the country. Unfortunately, in the ten years since, no active effort has been made to put in place mechanisms or policies to implement that principle.

In 1998 and 1999, the federal and provincial governments again engaged in extensive multistakeholder consultations. Not surprisingly, environmentalists again recommended large-scale change, based on the 'soft-path' energy vision which had been popularized by Amory Lovins. They argued that emissions could be cut 'to near half of 1990s levels through comprehensive building retrofits, a tripling of vehicle fuel efficiency, curbing urban sprawl, and the pervasive introduction of energy-efficiency and emission-reduction strategies into architecture, engineering design and urban planning' (Torrie and Parfett 2000: 24). No such vision was incorporated into national policy. Instead, the second federal-provincial national plan, released in 2000, again relied primarily upon voluntary action. Despite the Kyoto target, there had been no significant change in the policy dynamic which generated the 1995 national plan. That began to change, however, due to international pressure to make a yes or no decision on Kyoto ratification.

Although its economy is closely interlocked with that of the USA and the two countries had taken similar positions through the UNFCCC negotiations, Canada did not follow the American lead in 2001 when the Bush administration formally withdrew from the Kyoto regime. Instead, Canada took advantage of its increased bargaining leverage (specifically, the EU's willingness to grant concessions to keep Canada as a member and, thus, prevent the collapse of the regime) and pressed for a relaxation of its national objective by being allowed to count sinks as part of the national effort. At the July 2001 UNFCCC COP in Bonn, delegates agreed that Canada and other countries could count carbon stored in trees and soils (sinks) as part of their reduction effort. Chrétien, close to the end of his career and with an eye to the history books, then announced that the decision 'open[s] the way for... ratification by Canada in 2002' (Bjorn *et al.* 2002: 49). Those words initiated the most vociferous political battle yet seen in Canadian environmental policy.

The oil and gas industry and the province of Alberta engaged in media advertising campaigns intended to convince Canadians that ratification would cause major job losses. Several industrial sectors, including oil and gas, chemicals, electricity and motor vehicles came together to create the Canadian Coalition for Responsible Environmental Solutions, which also engaged in media advertising and lobbying of the federal government (Macdonald 2007). The business interests opposed to ratification made three major arguments. The first was that emissions reductions could only be achieved through technological developments which had to be phased in as the existing capital stock of buildings and machinery reached the end of its useful life, something which could not be achieved within the Kyoto timeframe. Secondly, business groups argued that Canadian policy had to be harmonized with that of the Bush administration to avoid a crippling competitive disadvantage because of the over-riding importance of exports to the American market. The third argument was that Canada should not spend public money on purchasing international credits, but should instead fund technology development at home (Macdonald 2003).

In one sense, the business campaign failed, since it did not prevent Canadian ratification. Furthermore, the federal government, once freed from the federal-provincial lowest-common denominator process which effectively gave Alberta a veto, abandoned voluntarism and announced that it would use law-based instruments to regulate industrial emissions. On the other hand, private negotiations resulted in an announcement by the Natural Resources Minister in December 2002, shortly after the ratification decision, that the oil and gas industry share would not exceed 15 per cent of total reductions, and that the cost of reductions by all sectors above C$15 per tonne would be paid by the federal government (Macdonald 2003). That policy of capping industry costs was reiterated in the Harper government policy announcement of April 2007.

Although they have abandoned voluntarism, the successive federal governments of Jean Chrétien, Paul Martin and Stephen Harper have not used regulatory instruments with a heavy hand. In 2002 the Chrétien government acceded to industry demands that regulatory standards be intensity based, governing the ratio of emissions to production, rather than impose absolute standards. This means that emissions can rise as production increases while still being in compliance with law, thus removing the Kyoto absolute cap from Canadian industry. Federal regulators privately negotiated the details of the regulatory programme with industry from 2003 until the Martin government fell in late 2005. By that date, no regulations had been put in place. In separate negotiations with the automobile industry, however, a voluntary agreement to change vehicle design to reduce emissions by 25 per cent had been agreed.

The Harper government, with its political support concentrated in western Canada, took office with an obvious inclination to do as little as possible on the issue. In autumn 2006 it announced a policy package which bundled together climate change and all other air pollution issues, and which for the former was essentially a return to voluntarism. At the time, however, Canada was chair of the UNFCCC COP, which meant that each international meeting resulted in considerable press attention to the gap between Canada's formal role and the implicit government policy of abandoning Kyoto. That publicity was coupled with mounting support for action in public opinion polls. On 26 January 2007, a Toronto newspaper reported a poll showing that when asked 'what is the most important issue facing Canada' 26 per cent of respondents cited 'environment' – beating the usual top-of-mind issues such as health, education or terrorism (Laghi 2007). The pollster said that: 'It's developed a top-of-mind salience the likes of which we've never seen before... In 30 years of tracking, we've never had over 20 per cent saying they think this is the most important issue' (Laghi 2007). This shift in public opinion led the Harper government to re-instate the previous government's regulatory programme. The April 2007 policy requires industry to reduce its emissions intensity by 18 per cent by 2010 or pay C$15 for each tonne of emissions over the regulatory limit (Environment Canada 2007). The government also intends to translate the voluntary 25 per cent motor vehicle reduction target into a regulatory programme.

The second policy instrument used by the federal government since ratification has been public spending, both as contributions to provincial programmes and for technology development. By the time the Martin government's spring 2005 budget was released, total federal government spending commitments totalled approximately C$10 billion, stretching through to the end of the 2012 Kyoto period (Canada 2007). The Harper government initially reduced spending in 2006 but, again pressured by public opinion, announced plans to spend $4.5 billion on environment in its 2007 budget, the bulk of it on climate (Simpson *et al.* 2007).

The three instruments of voluntarism, slow-moving intensity-based regulation and spending have been the only ones used to date by the Canadian government. A national emissions trading system has been studied, but no serious steps to implement it have yet been taken. No significant transportation policies, such as major funding increases for rail or urban transit, have been introduced, and no economic instruments, such as gasoline taxes to change incentives facing individual Canadians, have been used. Additionally, no attempt has been made to develop an explicit climate-energy policy. However, the Chrétien, Martin and Harper governments have all remained committed to maximum expansion of oil and gas exports to the USA, most notably from the Alberta oil-sands, while Harper has talked about Canada becoming an energy super-power. This basic conflict between energy policy and climate policy has been noted by a federal government watch-dog agency: 'First and foremost, the government needs to clearly state how it intends to reconcile the need to reduce greenhouse gas emissions against expected growth in the oil and gas sector' (Commissioner of Environment and Sustainable Development 2006: 12). No such statement has been made.

The only province to have introduced legal emissions requirements is Alberta, which, like the Harper government, has adopted a policy objective much weaker than the Kyoto goal and has used intensity-based standards. Some provincial governments have begun, in a hesitant manner, to introduce policies to increase land-use densities, support public transit and require that a portion of electricity be generated from renewable sources. However, a 2006 assessment by the David Suzuki Foundation (2006: 3) concluded that: 'Many provinces and territories still do not have climate change plans. Of those who do, few use the kinds of policies that have been shown to work elsewhere: regulations, a cap-and-trade system for greenhouse gases, financial disincentives like pollution fees and taxes.'

Throughout the period 1992–2007, elected leaders at the federal and provincial levels have worked to minimize the political cost of climate policy by favouring appearance over substance. The two periods of extensive and well-publicized multistakeholder consultation, 1993–1995 and 1998–2000, had the effect of delaying policy action and simultaneously giving an image of serious attention to the issue. The primary instruments used between 1995 and 2002, voluntary programmes for both industry and individual Canadians (the latter a *One-Tonne Challenge* presented primarily through television advertising), were also well publicized through annual awards and media advertising. Publicity is an essential component of voluntary programmes, since they rely on peer pressure rather than sanctions or financial incentives. For that reason, they are ideally suited for governments seeking to *appear* to be taking action. The decision to limit regulatory requirements to emissions intensity, rather than absolute caps was, conversely, relegated to the fine print of successive federal government plans. Politicians at all levels have talked at length about their commitment to action, while assiduously avoiding policy instruments which might be both effective and for that reason produce strong political resistance.

The policy dialogue respecting this policy failure

Academic and professional analysis the barriers blocking successful Canadian climate policy group into four categories: (i) the magnitude of the challenge; (ii) political opposition from energy-intensive industrial sectors and some provinces, most notably Alberta; (iii) the basic facts of both regionalism, with its associated differing economic interests, and the Canadian federal system of shared jurisdiction by the two senior levels of government; and (iv) inadequacies within the federal government administrative structure.

Magnitude of the challenge

Jeffrey Simpson, a political columnist with the influential Toronto newspaper, *The Globe and Mail*, has repeatedly noted the fact that both the Canadian population and the economy have experienced considerable growth since the early 1980s, exceeding that of climate-policy leaders such as Germany. In addition, he notes economic growth is based on energy-intensive industries, such as the western oil-sands, concluding that: 'No other G8 country faces these pressures'

(Simpson *et al.* 2007: 85). Similarly, Samson (2001), Toner *et al.* (2001) and Bjorn *et al.* (2002) point to the basic challenges of geography (long distances, cold winters, hot summers) and the importance of energy to the Canadian economy, while the Canadian Council of Chief Executives additionally stresses the effects of immigration and growing population (d'Aquino 2002–2003). Harrison and Sundstrom (2007) also document the fact that the Canadian Kyoto target represented a 29 per cent reduction of emissions below the 'business-as-usual' projection for 2010. The EU target on the other hand, due to the UK's switch from coal to gas and the closure of energy-inefficient plants in Germany after reunification, represents only a three to nine per cent cut below 2010 business-as-usual (Harrison and Sundstrom 2007).

The magnitude of the challenge has also been pointed out by the Canadian government. In 2007, the Harper government used cost arguments to justify its refusal to follow policy recommendations from the opposition parties which, they claimed, would have allowed Canada to meet the Kyoto goal by 2012. Although it made no comparisons with other Kyoto signatories, it argued that meeting the Kyoto target would require a 25 per cent rise in unemployment, a 50 per cent rise in the cost of electricity, a 60 per cent rise in gasoline prices, a C$4000 drop in annual disposable family income and a 6.5 per cent GDP decline, prompting 'a recession comparable to the one in 1981–1982, which stands as the largest recession to date in Canada since Second World War II' (Canada 2007: 2).

Political opposition

As noted, the announcement by Jean Chrétien in 2001 that his government planned to ratify the Kyoto Protocol provoked the largest political battle over environmental policy yet seen in Canada. Prior to that, business had consistently pressed publicly for only voluntary instruments to be deployed. After ratification and the federal government decision to move to regulatory instruments, it seems reasonable to assume industry has played an obstructionist role during subsequent private negotiations between 2003 and 2007. This business opposition has been noted by several commentators (Bernstein and Gore 2001; Macdonald 2003, 2007; Macdonald *et al.* 2004; Simpson *et al.* 2007).

Similarly, analysts have pointed to the blocking role played by Alberta. Smith (1998: 14) noted that: 'Clearly, Alberta has played, and will continue to play, a key role in climate policy development, as it protects its interests and those of the oil industry' (see also Macdonald 2003;

May 2003; Urquhart 2003). Rabe (2005) also points to the fact that the Alberta blocking role has not been countered by strong leadership from any other provinces, while Urquhart (2003: 25) makes the same point: 'genuine leadership on the Kyoto file, is nowhere to be seen in Canada's capital cities'.

Regionalism and federalism

Canada's population is spread in a long, thin line north of the US boundary and divided into four distinct regions: British Columbia on the Pacific coast; the West; the industrial heartland of Ontario and Quebec, with the latter being distinguished by separate language and culture; and Atlantic Canada. These regions have very different incentives in respect of energy and climate change. Alberta and Saskatchewan dominate Canadian oil and natural gas production, a considerable portion of which is exported to the USA. In recent years, the billions of dollars invested in the Alberta oil-sands have caused a booming provincial economy. Energy users, on the other hand, are concentrated in Ontario and Quebec, although Quebec meets the majority of its energy needs through relatively inexpensive hydro-electricity. Doern and Gattinger (2003: 25) note that: 'Canada has replicated within its borders some of the basic producer-consumer conflicts seen on the world stage between the Western OECD consumer countries and OPEC oil-producing countries'. Macdonald *et al.* (2004) highlight these differing regional economic interests as an important explanatory factor in Canadian climate policy, in particular the contrast between Quebec's willingness to act and Alberta's veto role.

The facts of regionalism and language identity led to the creation of Canada in 1867 as a federated state. In terms of whether the institutional scaffolding of Canadian federalism is strong enough to contain these regional energy and climate change interests, Winfield and Macdonald (2007) have compared two case studies of federal-provincial environmental policy. In the first, the system has been able to put in place harmonized environmental policy for toxic substances. With respect to climate, however, the system broke down completely when the federal government moved to ratify Kyoto in 2002. They explain this by pointing to differences in federal government motivation and the inherent weakness of the federal-provincial system, which could not accommodate the Alberta blocking role (also Harrison 2006; Meadowcroft 2007). Rabe (2005) has also noted that, throughout the federal-provincial process, provinces have spent more time and energy defending their own interests – by warding off stronger policy instruments which might damage provincial economies and by seeking federal funding – than in cooperating to develop effective national policy.

The fourth argument, that policy failure stemmed partially from administrative weaknesses in the Canadian government, has been made by a number of analysts. Smith (1998), Macdonald *et al.* (2004) and Simpson *et al.* (2007) all stress the conflict between the two lead federal departments – Environment Canada and Natural Resources Canada – as an impediment both to the development of effective federal policy and to the Canadian government's ability to coordinate provincial policies. Simpson *et al.* (2007) claim that the federal government's inability to manage inter-departmental conflicts (similar to those which exist in all governments between environmental departments and departments with mandates for industrial development) was due to Jean Chrétien's hands-off management style and willingness to leave files in the hands of his ministers. The federal Commissioner of Environment and Sustainable Development, a watch-dog federal agency located within the office of the Auditor-General, has also pointed to administrative inadequacies within the federal government (2006: 10).

> Our audits identified weaknesses in the government-wide system of accountability for climate change. Coordinating committees and mechanisms that once existed have been phased out and have not been replaced. A lack of central ownership, clearly defined departmental responsibilities, integrated strategies, and ongoing evaluation systems all point to problems in the government's management of the climate change initiative.

A possible fifth explanation, the reliance on voluntary policy instruments, and more particularly the VCR initiative, has been heavily criticized by environmentalists and academics. As Rivers and Jaccard (2005: 307) note: 'federal government climate change policy over the last decade has emphasized noncompulsory policies such as voluntarism, information provision, and modest subsidies. These policies are designed primarily to engender minimal political resistance, and have been relatively ineffective'. The choice of ineffective instruments is certainly a cause of policy failure but also a symptom of wider failures. In the following section I offer my own explanation for the choice of ineffective instruments and Canada's associated inability to achieve climate policy objectives.

Explaining policy failure

Given the history of Canadian climate policy, two failures in fact need explaining: that of the federal-provincial policy process which functioned from 1992 to 2002, and that of the unilateral federal and provincial policy-making which succeeded it. The first requires an understanding of why Alberta was strong enough to prevent any instrument other than voluntarism being used. The second requires exploration of why the federal government and pro-Kyoto provinces such as Quebec, once acting alone and free of the Alberta blocking role in the federal-provincial decision-making process, have remained reluctant to use stronger instruments. I first assess each by discussing the adequacy of the four explanatory factors reviewed above.

The fact that geography and growth of the economy and population have meant that Canada faces a larger challenge than other Annex 1 countries is, arguably, irrelevant because Canadian policy, both national and unilateral federal, has to all intents and purposes had *no* impact on annual increases in emissions. The magnitude of the challenge might explain partial policy failure but cannot explain the complete failure which has flowed from the decision to use ineffective instruments.

There is stronger evidence that two of the other explanations given above, federalism and inadequacies in the federal government policy process, constitute part of the explanation. By themselves, however, they cannot provide a full answer. The Canadian constitution gives the federal government all the powers necessary to make more effective national policy, including disallowance of provincial laws. Such powers cannot be exercised, however, in a country continually facing the threat of provincial separatism. The de facto federal-provincial system, accordingly, is very weak, roughly analogous to international regimes, which are hampered by the right of sovereign states to opt out of any agreement. However, even this system could have generated stronger policy had two other conditions been different: first, an engaged and determined federal government and, second, at least one province committed to strong action on climate change and prepared to play the role California traditionally has with respect to US air pollution policy. In the same way, a committed Prime Minister could have used constitutional powers to solve the problem of feuding federal departments. Accordingly, we must look not only to the system but also to another factor, the *lack of commitment and political will* on the part of the Chrétien, Martin and Harper governments and also among all Canadian provinces.

This lack of political will is directly related to the fourth factor discussed above, opposition by business and some provinces. Although this provides a large part of the explanation, the political strength of those opposing a policy measure cannot be considered in isolation but instead must be considered *relative* to that of those supporting the measure. In this case, opposition to stronger climate policy measures was able to succeed because of the absence of politically strong support from either public opinion or the environmentalists for whom such opinion is the major source of political power. Essentially, the political weakness of the Canadian environmental movement is also a key explanation for the lack of commitment in Ottawa or any of the provincial capitals.

Accordingly, I argue that to understand the failure of both national and federal government policy we must look, first, to the lack of federal government leadership and, second, to the lack of a provincial champion to support federal leadership. Both factors in turn rest upon a third: the relative weakness of Canadian environmentalism and its inability to counter business and provincial opposition. I discuss each briefly and then conclude by suggesting the primary action required of a committed federal government.

Part of the explanation for federal government inaction can be found in the ideologies of the governing Liberal and Conservative parties. The former is a centrist party which has governed for a large part of the twentieth century by moulding its principles to fit current political fashions. Lacking any strong commitment to environmentalism, it had a relatively poor record during its time in office between 1993 and 2005. Prime Minister Chrétien intervened to push for stronger climate policy in 1997 and again in 2002, but otherwise ignored the issue. Harper leads a party which pays marginally more attention to core principles, but these tend to favour markets over environment. Despite this, the record shows that both parties have responded to pressure from environmentalists. The Liberals ran on a strong pro-environment platform in 1993, in the aftermath of the Rio conference, when popular support for environmentalism was just beginning to ebb. The Conservative party responded to public opinion in the spring of 2007, strengthening the climate policy it had announced the previous year. Had the pressure which influenced policy in 1993 and again in 2007 been sustained during the intervening years, it is fair to speculate that ideology would not have prevented either party from acting more decisively.

In terms of provincial champions, Ontario has traditionally stood shoulder to shoulder with the federal government on major national issues. In this case, however, Ontario was governed from 1995 to 2003 by the Mike Harris and Ernie Eves neoliberal governments, which explicitly took their cue from the Reagan US and Thatcher UK administrations. They were replaced by the current Liberal government which has a somewhat stronger environmental record, but which is hampered on climate change by the importance of the motor vehicle industry to the Ontario economy. Nor has the inward-looking Parti Québécois government, in power for most of this period with a mandate to lead the province out of confederation, had any interest in leading policy development at national level.

Finally, there is the question of why the Canadian environmental movement has lacked the political power necessary to force provincial or federal governments to put in place stronger policy. The answer does not appear to lie in differences in agency power. No empirical studies have been completed of the financial and staffing resources committed by the two camps to lobbying on climate policy since the early 1990s. However, personal communications indicate that, aside from the oil and gas industry spending millions of dollars on anti-ratification public advertising in 2002, neither camp has had more than ten to twenty people working full time on the issue (personal communication 2003).

This rough parity in agency power, however, does not carry over to structural power. Two aspects of the Canadian institutional structure limit the power of environmentalists relative to business. The first is the Westminster-style electoral system, which disadvantages marginal parties whose strength is not geographically concentrated. The Green Party has achieved nearly ten per cent of the popular vote in recent elections, but the geographic dispersal of this support has prevented it from electing a single member to the federal House of Commons or a provincial legislature.

Secondly, the environmental policy-making process guarantees private, elite-level access to business and for the most part denies it to environmentalists, since the essence of Canadian environmental regulation is private regulator–firm negotiation. The results of that structural advantage are evident in the history of Canadian climate policy-making. The one climate-policy battle which business has lost was the 2002 ratification decision. When the issue lacked salience prior to that, business was able to convince governments to rely on voluntary measures and since then, in private negotiations, it has been able to ensure that climate regulation remained relatively toothless. The visibility of the ratification process mobilized a number of supporters beyond the environmental movements, such as labour, health and backbench federal Liberal MPs. Business success since then, after the policy process reverted to private negotiation, confirms Smith's finding (2000) that business is more likely to win policy battles in private than in public.

Beyond these institutional weaknesses, Canadian environmentalism suffers from the fact that public opinion has been placated by the government strategy of giving the appearance of action, while avoiding the substance. As far as we know, the Harper government policy announcement of April 2007 resolved its political problem, despite the weakness of the regulatory regime announced and the failure to commit to the Kyoto target. For all these reasons, environmentalists have been unable to counter business opposition to stronger climate policy.

Conclusion

To summarize, Canadian climate-change policy-making during the 20-year period starting with the 1988 Toronto conference has been characterized by brief bursts of policy activism, interpolating a consistent refusal by federal and provincial policy-makers to engage seriously with the issue. The former occurred at the times of the 1988 Toronto conference; support for the UNFCCC at Rio in 1992 followed by early ratification; willingness to accept a challenging policy objective at Kyoto in 1997; ratification in 2002, in the face of considerable opposition; and the policy reversal by the Harper government in spring 2007. Between those intervals, policy-makers have given the appearance of action, through consultation, study and successive action plans, but have consistently avoided introducing more coercive policy instruments. This is because, with the exceptions of those brief moments, opponents of meaningful policy have generally been able to muster more power to influence to policy than have environmentalists.

Given this history, two things are needed for Canada to put in place stronger climate policy. The first is more powerful pressure to counteract the business lobby and prod Canadian governments, particularly the federal government, into action. Second, once so motivated, the federal government will need new strategies to address the political barriers which to date have blocked more radical action.

In terms of the first, pressure has to come from two sources – externally, from other states, most notably those of the EU, and from the governance institutions of the UNFCCC – and internally, from Canadian environmentalists. Canadian Prime Ministers have responded to pressure from other heads of state in the past, making it reasonable to predict that they will do so again. Pressure from the UNFCCC governing institutions can only come in the form of increased auditing and public reporting on Canada's performance. This in turn will depend on increased financial and staffing resources for environmental regime governance, something badly needed but beyond the control of any one country. Environmentalists within the country, however, could do more than they have to date to raise the visibility of climate policy during periods between bursts of government activity, when Canadians return to assuming that the issue is well in hand. They could divert resources from other issues to focus lobbying on climate; they could create a new, high-profile pan-Canadian coalition of groups pressing for climate action (churches, labour unions, cities, renewable energy industries, health groups and Arctic communities suffering climate impacts); they could more actively recruit champions within the world of finance and industry, much as George Soros does on global capitalism; and they could do more to tailor their message to new Canadians, immigrants speaking different languages and with different world views, who are an untapped resource. Above all, they should publish more policy analysis to show Canadians that the appearance of policy action created by successive governments is not matched by the reality. As noted, although recent polls show high public concern for the issue, the Harper government was still able to appease this while taking minimal policy action.

If such actions were to succeed, a motivated federal government would still have to surmount the basic political challenge stemming from Canada's federal governance structure and the fact that provinces like Alberta have both constitutional jurisdiction and strong economic incentives to oppose action. Here, the federal government must adopt a new strategy. As we have seen, it initially engaged in national lowest-common denominator decision-making, with the result that Alberta's veto power guaranteed continued reliance on voluntary instruments. Once that process ended in 2002, the federal government ignored Alberta while attempting unilaterally to regulate large industrial emitters. To achieve national success, however, the federal government must bargain with, not ignore, opposing provincial interests. The federal government has to actualize the principle enunciated in 1997 that no region should bear an undue portion of the total national cost. This means that, perverse as it may sound, one of Canada's wealthiest provinces will have to be compensated.

There are several ways in which this might be done. The federal government might coordinate a process whereby the provinces publicly bargain over their share of total emission reductions, much in the way UNFCCC parties bargained at Kyoto. Instead of all provinces cutting emissions by the same relative amount, recognition would be given to Alberta's 'national circumstances'. Beyond that, the federal government might increase subsidies for technological development in Alberta industries, ideally financed by some form of carbon tax. The Harper government is already putting in place a technology development fund financed by carbon charges. Explicitly deciding that a portion of those revenues would flow to Alberta industries might help to bring the province back into the national policy process. Whether through these or other measures, the federal government has to send a tough-love message to Alberta and the other oil-producing provinces that it understands their position and will assist, but will no longer turn a blind eye to their inaction.

References

Bernstein, S. and C. Gore (2001), 'Policy implications of the Kyoto Protocol for Canada', *Isuma: Canadian Journal of Policy Research* 2 (4), 26–36.

Bjorn, A., S. Duminuco, J. Etcheverry, C. Gore, D. Harvey, I. Lee, D. Macdonald, K. MacIntyre, R. Riddle, D. Scott, T. Soltay, K. Stewart and T. Zupancic (2002), *Ratification of the Kyoto Protocol: A Citizen's Guide to the Canadian Climate Change Policy Process*, Toronto: Sustainable Toronto.

Canada (2007), *The Cost of Bill C-288 to Canadian Families and Business*, Ottawa: Her Majesty the Queen in Right of Canada.

Commissioner of Environment and Sustainable Development (2006), *Report of the Commissioner of Environment and Sustainable Development to the House of Commons: The Commissioner's Perspective – 2006; Climate Change: An Overview; Main Points*, Ottawa: Minister of Public Works and Government Services.

Corcoran, T. (1994), 'Goodbye carbon tax, hello sanity', *The Globe and Mail, Report on Business*, Toronto.

d'Aquino, T. (2002–2003), 'A business case for responsible climate change', *Policy Options* 24, 50–2.

David Suzuki Foundation (2006), *All Over the Map: 2006 Status Report of Provincial Climate Change Plans*, Vancouver: David Suzuki Foundation.

Doern, G. B. and M. Gattinger (2003), *Power Switch: Energy Regulatory Governance in the Twenty-First Century*, Toronto: University of Toronto Press.

Environment Canada (2007), 'Action on climate change and air pollution', *News Release*, 26 April.

Harrison, K. (2006), 'The struggle of ideas and self-interest: Canada's ratification and implementation of the Kyoto Protocol', *Paper to the Annual Meeting of the International Studies Association*, San Diego, California.

Harrison, K. and L. Sundstrom (2007), 'The comparative politics of climate change', *Global Environmental Politics* 7 (4), 1–18.

Laghi, B. (2007), 'Climate concerns now top security and health', *The Globe and Mail*, Toronto, 26 January.

Macdonald, D. (2003), 'The business campaign to prevent Kyoto ratification', *Paper to the Annual Meeting of the Canadian Political Science Association*, Dalhousie University.
Macdonald, D. (2007), *Business and Environmental Politics in Canada*, Peterborough: Broadview Press.
Macdonald, D., A. Bjorn and D. VanNijnatten (2004), 'Implementing Kyoto: when spending is not enough', in Doern, G. (ed.), *How Ottawa Spends: 2004–05*, Montreal: McGill-Queen's University Press, p. 184.
May, E. (December 2002–January 2003), 'From Montreal to Kyoto, how we got from here to there – or not', *Policy Options* 24, 14–18.
Meadowcroft, J. (2007), 'Building the environmental state', *Alternatives Journal* 35, 11–17.
Pembina Institute (2007), *Analysis of the Government of Canada's April 2007 Greenhouse Gas Policy Announcement*, http://pubs.pembina.org/reports/Reg_framework_comments.pdf [4 February 2008].
Rabe, B. (2005), 'Moral super-power or policy laggard? Translating Kyoto Protocol ratification into federal and provincial climate policy in Canada', *Paper to the Annual Meeting of the Canadian Political Science Association*, London, Ontario.
Rivers, N. and M. Jaccard (2005), 'Canada's efforts towards greenhouse gas emission reduction: a case study on the limits of voluntary action and subsidies', *International Journal of Global Energy Issues* 23, 307–23.
Samson, P. (2001), 'Canadian circumstances: the evolution of Canada's climate change policy', *Energy and Environment* 12, 199–215.
Simpson, J., M. Jaccard and N. Rivers (2007), *Hot Air: Meeting Canada's Climate Change Challenge*, Toronto: McClelland and Stewart.
Smith, H. (1998), 'Stopped cold', *Alternatives Journal* 24 (4), 10–16.
Smith, M. (2000), *American Business and Political Power: Public Opinion, Elections and Democracy*, Chicago: University of Chicago Press.
Toner, G., D. Russell and B. Masterson (2001), 'Science and policy (im)mobilization: climate change policy in Canada', unpublished paper.
Torrie, R. and R. Parfett (2000), 'Minding the gap', *Alternatives Journal* 26 (2), 22–9.
Urquhart, I. (2003), 'Kyoto and the absence of leadership in Canada's capitals', *Policy Options* 24, 23–6.
Winfield, M. and D. Macdonald (2007), 'The harmonization accord and climate change policy: two case studies in federal-provincial environmental policy', in Bakvis H. and G. Skogstad (eds), *Canadian Federalism: Performance, Effectiveness and Legitimacy*, Don Mills: Oxford University Press, pp. 1–22.

13
Climate Policy in the USA: State and Regional Leadership

Allison M. Chatrchyan and Pamela M. Doughman

Introduction

As Brewer and Pease noted in Chapter 5, several obstacles continue to inhibit the enactment of comprehensive federal climate change policies in the USA, including divisions within government and an increasing polarization of politics; well-entrenched interests that favour the status quo; and low public awareness or interest in addressing climate change. If the analysis of US climate politics ended there, the prospects would look dire indeed. Luckily, developments at sub-national level suggest a more hopeful picture, with significant policies being adopted to reduce greenhouse gas emissions at the state, regional and local levels. Rabe (2006: 18) observes that 'the burgeoning state role must be seen as not merely an extension of existing authority but rather a new movement... driven by a set of factors distinct to the issue of climate change', including climate impacts, economic development and advocacy coalitions for strong climate policy action within key arms of government.

Building on Rabe's observations, this chapter argues that the climate policies of California, New York and other states and regional initiatives represent a new kind and depth of leadership. This new state leadership is made possible by the federalist nature of US politics and key states' willingness to embrace the idea that becoming climate leaders is in their economic and environmental interests. These state climate leaders hold very different conceptions of climate science, economic considerations and ethical perspectives than the G. W. Bush Administration and have been willing to act on those convictions. As of July 2007, the population of California and the ten Northeastern states was almost 85.3 million, 28.3 per cent of the US population (US Census Bureau

2007). Their combined CO_2 emissions from fossil fuel combustion alone in 2004 were almost 1028 million tonnes of CO_2 ($MtCO_2$), more than 18 per cent of total US CO_2 emissions (US Environmental Protection Agency [US EPA] 2007a; 2007b). For comparison, Germany's total CO_2 emissions in 2004, including emissions from fossil-fuel combustion and other sources, were about 897 $MtCO_2$, less than California and the ten Northeastern states combined (United Nations Framework Convention on Climate Change 2007). Clearly, the greenhouse gas reductions of these states are important in a national and a global context.

Climate policies at the state level

Federal efforts to reduce greenhouse gas emissions intensity in the USA have so far failed to curb the overall growth in US emissions because they have not been coupled with mandatory reductions, strong technology development incentives and market mechanisms (COG 2006a). As Brewer and Pease noted in Chapter 5, of the 50 major climate policies summarized in President Bush's *Climate Action Report* of 2002, only six were regulatory; the rest were voluntary.

Faced with inadequate action at the federal level, several US states began adopting their own policy measures on climate change in the late 1990s. While climate leadership at the state level in the late 1990s and early 2000s was largely limited to California, New York and a few other early adopters, by 2007–2008 the tide had begun changing, with the majority of the 50 US states beginning to have some climate policies in place. This not only includes the traditionally liberal-democratic states of California and the Northeast but now includes some of the most conservative states of America's heartland. According to the National Caucus of Environmental Legislators (NCEL 2007), 39 states considered 349 climate change proposals in their legislatures in 2007 (Billings 2007). US states have adopted numerous climate policies, including requirements for the use of renewable energy and mandatory greenhouse gas emissions reduction targets (see Table 13.1).

California's climate change policies

California has had longstanding policies supporting energy efficiency, pollution reduction and renewable energy. In the late 1990s, these were portrayed as 'no regrets' policies because they were beneficial to the economy for non-climate-related purposes and also set California on

Table 13.1 State policies on climate change, as of March 2008

Climate policies	**State action**
Active climate legislative commissions or executive branch advisory groups	23 states including California and New York
Greenhouse gas emissions targets	17 states adopted, including California AB32, Maine, New York
States adopting California's *Automobile Emissions Standards* for cars and light trucks, which will include a greenhouse gas emissions standard with model year 2009	14 states adopted or are in the process of adopting
Renewable Portfolio Standard (RPS) requiring energy companies to procure a certain percentage of their power from renewable resources by a certain date	24 states adopted in some form
Regional cap-and-trade programmes	21 states participating in WCI (7 and British Columbia, Manitoba), RGGI (10) or Midwestern Regional Greenhouse Gas Reduction Accord (6 states plus Manitoba). Observers: RGGI (1), WCI (6 states, 3 Canadian provinces, 4 Mexican states), Midwestern (3)

Source: NCEL (2007); Western Climate Initiative (WCI) (2008a; 2008b).

a path to respond to growing concerns about the effects of human-caused greenhouse gas emissions (CEC 1998). Since 2000, California has adopted numerous additional climate policies, the most significant of which are described below.

In 2001, the California Climate Action Registry (CCAR) was established by California statute as a non-profit public–private partnership to record and register voluntary greenhouse gas emissions reductions made after 1990. As of January 2008, 319 organizations were participating in the programme (CCAR 2007). *Law AB1493*, enacted in 2002, requires personal vehicles in model year 2009 and later sold in California to meet a greenhouse gas emission standard (OCLI 2002). This law has drawn recent national media attention because its implementation requires US Environmental Protection Agency (US EPA) approval of a waiver for California under the Clean Air Act. If a waiver is granted, other states may adopt the federal standard or the California standard. In December 2007, the US EPA denied the waiver, which prompted California, 15 other states and five environmental groups to file a suit against the US EPA (Barringer 2008). At least 16 states are poised to adopt California's vehicle greenhouse gas emissions standard if the waiver is granted (Broder and Barringer, 2007). In June 2005, California Governor Arnold Schwarzenegger, a Republican, issued *Executive Order S-3-05*, which set greenhouse gas targets for California to reach 2000 emissions levels by 2010; 1990 emissions levels by 2020; and an 80 per cent reduction below 1990 levels by 2050. The Executive Order also created the California Climate Action Team to implement emissions reduction programmes and report on progress.

Perhaps most important and well known of California's legislation is AB 32, the Global Warming Solutions Act, which Governor Schwarzenegger signed into law in September 2006. AB 32 puts into law the Governor's second target of reducing greenhouse gas emissions to 1990 levels by 2020 and states that California intends to continue reducing emissions after 2020. AB 32 also directs the California Air Resources Board to adopt regulations to achieve the maximum technologically feasible and cost-effective emissions reductions in support of achieving the statewide greenhouse gas emissions limit. It is also the first state programme in the USA to mandate an economy-wide emissions cap with enforceable penalties (Pew Center 2007a). AB 32 addresses the role of sub-national action on this global environmental problem as follows: 'National and international actions are necessary to fully address the issue of global warming. However, action taken by California to reduce emissions of greenhouse gases will have far-reaching effects by encouraging other states, the federal government, and other countries to act' (OCLI 2006).

In January 2007, the Governor also issued *Executive Order S-01-07*, establishing a target to reduce the carbon intensity of California's transportation fuels by at least ten per cent by 2020 and directing the California Air Resources Board to develop a Low Carbon Fuel Standard for transportation fuels in California. The Governor also urged the Bush administration to do more to address climate change (COG 2007):

> When it comes to energy and alternative fuels, California is setting the standard for reducing our dependence on fossil fuel by establishing the first Low-Carbon Fuel Standard. Like my order on alternative fuel, I believe it is critical that any policy from the President or Congress has teeth and is not just a noble concept.

Climate change policies in New York

Since the late 1990s, New York and California have almost been in a friendly competition to be the strongest state leader on climate change. New York has longstanding policies supporting energy efficiency, pollution reduction and renewable energy. In 1996, Governor Pataki, also a Republican, proposed the Clean Water/Clean Air Bond Act, approved by New York voters in 1996, to include $55 million for clean-fuelled vehicles and buses, and worked to promote the biofuels industry. New York established a System Benefit Charge on the sale of electricity to support energy efficiency, renewable energy and environmental research in 1998, and established the NY Energy Smart Program to be administered by the New York State Energy Research and Development Authority (NYSERDA 2008). The Governor established a Greenhouse Gas Task Force in 2001 to assist in developing policy recommendations and strategies to reduce New York's emissions. Also that year, Governor Pataki issued Executive Order 111, *Green and Clean State Buildings and Vehicles Guidelines,* which requires the state government to achieve, by 2010, a reduction in energy use of 35 per cent relative to 1990 levels and the purchasing of 20 per cent of its energy from renewable energy sources (CACP 2005).

New York's comprehensive State Energy Plan was adopted in 2002 and set a greenhouse gas emissions reduction goal of five per cent below 1990 levels by 2010 and ten per cent below 1990 levels by 2020 (CACP 2005). Other programmes adopted under Governor Pataki include a state tax credit passed in 2000 for the construction and rehabilitation of 'green buildings'; a renewable portfolio standard requiring 25 per cent clean energy by 2013; and adoption of California's zero emission vehicle rule (CACP 2005). In 2005, Governor Pataki signed Executive Order No. 142 *Directing State Agencies and Authorities to Diversify Fuel and Heating Oil Supplies Through the Use of Biofuels in State Vehicles and Buildings* (State of New York Executive Chamber 2005).

New York also initiated three programmes in the transportation and energy efficiency sectors: the Clean-Fueled Bus Program; the Advanced Travel Center Electrification Program for use by truckers along New York State highways to reduce idling, fuel consumption and emissions; and the Keep Cool Air Conditioner Bounty Program to reduce peak demand for electricity (Pew Center 2008). New York State also provided leadership in the initial development and backing of the Regional Greenhouse Gas Initiative (RGGI), which is discussed further below.

New York is continuing to move ahead with progressive climate change policies. In April 2007, former New York Governor Spitzer announced the most aggressive plan of any state to reduce global warming, while addressing the economic problems resulting from rising energy prices in New York, which now are the second highest in the nation behind Hawaii. The *15 by 15* plan is projected to cut the state's electrical consumption by 15 per cent from levels forecasted for 2015, while building enough clean generating capacity to lower power costs (Iwanowicz 2008). Spitzer also created a more robust Climate Change Office within the New York State Department of Environmental Conservation (DEC), with a dedicated staff of 12 scientists, economists and policy analysts, which is critically important, since the state 'is on the verge of creating the United States' first greenhouse gas cap-and-trade system' through the RGGI (Moore, quoted in Klein 2007). These climate change policies and programmes are being continued under New York Governor David Paterson, a Democrat sworn in to office in March 2008 upon Spitzer's resignation.

Regional climate initiatives in the USA

In addition to activities within their territories, California and New York have been instrumental in encouraging neighbouring states to start addressing climate change through regional initiatives. By the end of 2007, 23 states were participating in regional cap-and-trade programmes, with nine states participating as observers (see Figure 13.1). Regional climate change initiatives are more efficient than individual state programmes since they eliminate duplication of efforts, include a broader geographical area and create more uniform regulations (Pew Center 2007b).

In 2003, New York's Governor Pataki invited 11 state governors from the Northeast to Mid-Atlantic region to participate in a regional programme to reduce greenhouse gas emissions from power plants (RGGI 2008a). The RGGI now includes 10 states from Maine to Maryland. It was the first regional group to agree to develop a mandatory cap-and-trade programme in the USA, capping emissions at current levels from 2009 to 2014, with a ten per cent reduction by 2019 (RGGI 2007a). To guide participants' implementation of the programme, RGGI issued a Model Rule in August 2006 (RGGI 2008b). A critical aspect of the RGGI cap-and-trade programme is that emissions allowances will be auctioned off, rather than allocated for free, in contrast to the first two phases of the European Union's emissions trading scheme. The sale of allowances will generate funds for projects that reduce greenhouse gas emissions further (DEC 2007; 2008; RGGI 2007a).

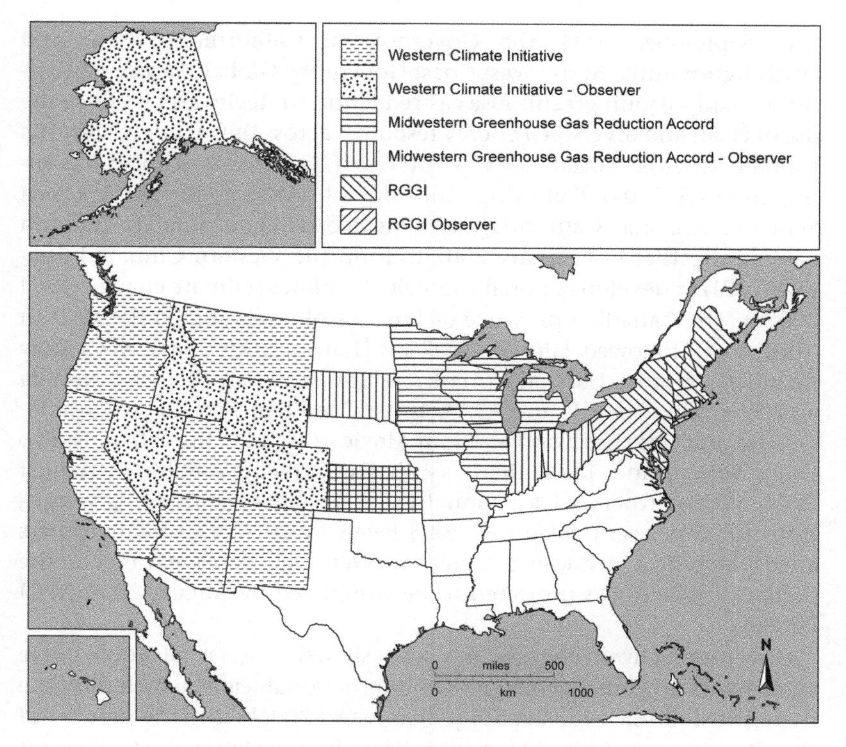

Figure 13.1 Regional initiatives to address climate change in the USA
Source: Pew Center (2007a).

RGGI states have agreed to participate in regional auctions of CO_2 allowances on a quarterly basis, and also have agreed to a number of design elements for the mandatory cap-and-trade-system. In March 2008, the participating states announced that the first RGGI allowance auctions will be held in September and December 2008, while the first compliance period for the programme will begin in January 2009. In his announcement of this milestone, Pete Grannis, the NY DEC Commissioner, noted, 'Climate change is the most significant environmental problem of our generation.' He also said, 'Absent federal leadership, the Northeast and Mid-Atlantic states of RGGI are taking action to cut greenhouse gas emissions and reduce their impact on the environment. Our CO_2 auction will be the first in the nation and should be replicated at the federal level' (RGGI 2008c).

In September 2003, the Governors of California, Oregon and Washington initiated the West Coast Governors' Global Warming Initiative to explore joint greenhouse gas-reduction strategies and increase the use of clean and diversified energy resources across the region (California Climate Change Portal 2004; West Coast Governors' Global Warming Initiative 2004). Following California's leadership, the five Western States of Arizona, California, New Mexico, Oregon and Washington banded together in February 2007 to form the Western Climate Initiative (WCI) to develop regional strategies to address climate change (WCI 2008a). The Canadian province of British Columbia joined the WCI in April 2007, followed later in 2007 by Utah and Montana. WCI now includes six more US states (Alaska, Colorado, Idaho, Kansas, Nevada and Wyoming) (see Figure 13.1), three Canadian Provinces (Ontario, Quebec and Saskatchewan) and four Mexican states (Chihuahua, Nuevo Leon, Sonora and Tamaulipas) as observers (WCI 2008b). In August 2007, WCI partners set a regional target to cut greenhouse gas emissions to 15 per cent cut below 2005 levels by 2020. The WCI partners are developing a market-based mechanism to help achieve this goal, the design of which was projected to be completed by August 2008 (WCI 2008a).

Governor Schwarzenegger has also signed a series of cooperative agreements on climate change with international leaders, including the then British Prime Minister Tony Blair (COG 2006b) and the premier of the Canadian province of Manitoba. Regarding the latter, the Governor's press release explained:

> The science is clear. The global warming debate is over. We have a responsibility to act decisively to slow or even stop climate change. But California cannot do it alone. Reducing greenhouse gas emissions is a global effort.
>
> (COG 2006c)

Until late 2007, the states involved in developing regional climate initiatives were the more liberal states on the Eastern and Western coasts. However, in November 2007, more conservative states in the heartland of the USA established an agreement to reduce emissions. Illinois, Iowa, Kansas, Michigan, Minnesota and Wisconsin, along with the Canadian Province of Manitoba and three other observer states (Indiana, Ohio and South Dakota), established the third major regional climate initiative in the USA, the Midwestern Regional Greenhouse Gas Reduction Accord. Its members agreed to establish regional greenhouse gas reduction targets within 12 months, including a long-term target to reduce emissions to 60–80 per cent below current levels, and to develop a multi-sector cap-and-trade system to help meet the targets (Pew Center 2007a; Office of the Governor, Wisconsin 2007).

At the local level, a multitude of communities throughout the USA are participating with the International Council for Local Environmental Initiatives (ICLEI) Cities for Climate Protection Program, which requires participating local governments to (i) establish greenhouse gas baselines; (ii) set targets; (iii) develop local action plans; (iv) implement action plans and (v) measure results. By October 2007, 28 towns and counties in New York had signed up to the programme, while 78 cities, counties and water districts in California had adopted it (ICLEI 2008). More than 800 municipalities in the USA have also signed the US Mayors' Climate Protection Agreement, an effort led by the Seattle Mayor Greg Nickels. This phenomenal degree of local action attests to a growing energy of grass-roots, volunteer-led efforts within local civil society for 'bottom-up' action to address climate change in response to the absence of federal climate change leadership.

Further greenhouse gas emissions reductions are needed

As Barker noted in Chapter 1, countries at the G8 meeting in 2007 agreed on the need to reduce global greenhouse gas emissions by at least 50 per cent below 1990 levels by 2050, while Intergovernmental Panel on Climate Change reports (IPCC 2007) state that CO_2 emissions reductions of 50–85 per cent from 2000 emissions will be required to limit the global temperature rise to 2.0–2.4 °C above pre-industrial levels. Although the near-term regional, state and local initiatives described above would, if achieved, result in significant emissions reductions, it is clear that further reductions are needed. Near-term measures should, thus, be designed with a clear eye on facilitating deeper reductions in the future (Schiller 2007). Current RGGI and WCI commitments do not extend beyond 2020; however, a number of states and regional initiatives have adopted longer term reduction goals that are consistent with, or exceed, the level of global emissions reductions recommended by the G8 and IPCC. In most cases, however, the post-2020 targets are not yet incorporated into state law (Pew Center 2007c).

Why are states taking the lead on climate change?

While the political window of opportunity has not yet opened for comprehensive, national climate change legislation in the USA, it is apparent from the foregoing review that state and regional initiatives have been able to mobilize much greater political support for stronger climate policies. A key explanation for this is the USA's federal system of government, where sovereignty is divided between the central governing authority and constituent political units. Within this system, states have the authority to address climate change under the tenth amendment of the US Constitution, which affirms that powers not delegated to the USA by the Constitution, nor prohibited by it to the states, are reserved to the states respectively, or to the people.

Regarding environmental policy, federalism provides the legal foundation for states to experiment with policy solutions when the federal government is inactive or insufficiently active. It also allows states to tailor policy design and resources to meet local priorities and conditions. Those who favour environmental policy devolution to the states argue that local decision-makers better understand local issues, have better local data and can use flexible, innovative methods that might not work with a national approach (Rothenburg 2002). In addition, the Pew Center reports that: 'states have primary jurisdiction over many areas – such as electric generation, agriculture, and land use – that are critical to addressing climate change' (Pew Center 2007d: 1). Policy devolution in this case also makes sense because climate change is expected to have important, but uneven, impacts as a consequence of states' differing geographies (e.g. coastal and interior states versus mountainous and plains states).

In 2002, the Pew Center released the following summary of a study by Barry Rabe on US climate polices at the state level (Pew Center 2002; Rabe 2002):

> ... [S]tate actions on climate change play a unique role in overall climate change governance. The United States comprises diverse regions, and individual state policies can be tailored to each state's strengths. State policies have the potential to spread among states, as is occurring already with several of the programs featured in the report. Successful state actions can become prototypes for federal programs.

In the years since this report, many additional states have adopted climate policies and committed to regional initiatives. Equally, although states have designed innovative environmental policies in the past, state policies on climate change since 2000 show a new type of leadership, emphasizing cooperation among states and across borders through regional climate change agreements. Historically, some states have shown leadership in adopting environmental standards more stringent than federal standards. For example, in developing its State Implementation Plan under the Clean Air Act, California showed important leadership by adopting standards more stringent than required by the Act. However, these standards were adopted within the context of a federal law that was already quite ambitious. In the case of climate change, there is no ambitious federal law in place but the United States' federal structure has enabled policy action at the local, state and international levels.

If federalism has been a key catalyst for states and regional initiatives, and for states to apply pressure for federal government action, a core reason for state governments to adopt stronger climate policies is the desire to reduce the damages and costs of mitigation and adaptation. This has been driven in no small part by state governments' interpretations of scientific data on the observed and expected future impacts of climate changes. The California EPA (CEPA 2005) reports that La Jolla has seen about a 15 cm rise in sea level between 1900 and 2004. In addition, average California spring snowmelt has decreased 12 per cent between 1906 and 2004. Both these factors could have major impacts on freshwater supplies and infrastructure. States in the Northeast are also aware that their climate is changing in noticeable ways. According to an October 2006 report from the Northeast Climate Impacts Assessment (2006: 1): 'temperatures have been rising, particularly in winter, and the number of extremely hot days in summer has been increasing'. States have, thus, been willing to take early action on climate change because they perceive that reducing the amount of climate change also reduces adaptation costs, while early investment in low carbon technologies can stimulate state economies, create jobs and other first-mover advantages, improve local air quality, and reduce exposure to volatile energy prices (Pew Center 2007d).

Whatever the motivations of individual state governments, such state leadership on climate change demonstrates the value of the 'new federalism' to environmental policy in the USA: states may be better attuned to their local constituencies and better able to adopt and hone difficult but necessary environmental mandates that are not yet politically feasible at the federal level. For example, there are widely disparate public conceptions and communications about climate science and policy across the country. At a national level, the issue remains a low priority for the average person (behind the economy, education and the Iraq war), and a number of politicians have portrayed climate science as insufficiently certain to require mandatory greenhouse gas reductions (White House 2001). The G. W. Bush administration has also portrayed actions to reduce greenhouse gas emissions as imposing unacceptable costs on the economy and has highlighted the negative effects of policies to curb emissions on deserving groups such as working families:

> For America, complying with those mandates would have a negative economic impact, with layoffs of workers and price increases for consumers. And when you evaluate all these flaws, most reasonable people will understand that it's not sound public policy.
>
> (White House 2001)

In California, on the other hand, there is greater public awareness of the issue and more widespread support for state policy to address climate change. A Public Policy Institute of California poll in July 2007 found that:

> [L]arge and growing proportions of residents are in favor of California making its own policies, separate from the federal government, to address global warming (54 per cent July 2005, 65 per cent July 2006, 67 per cent today).... Reflecting their strong support for state-level efforts, more than eight in 10 Californians... favor the 2002 state law that requires all automakers to further reduce greenhouse gas emissions from new cars in California beginning in 2009. Favor for this policy has been overwhelming among California adult residents each time we asked about it over the past five years (81 per cent June 2002, 80 per cent July 2003, 81 per cent July 2004, 77 per cent July 2005, 78 per cent July 2006, 84 per cent today).
>
> (Baldassare *et al.* 2007: 11)

In California and the Northeastern states, greenhouse gas emissions reduction is viewed, as well as an issue of public concern, as a business opportunity for technological innovation to avoid harm to a broad cross-section of environmental, economic and social interests. Hoffman (2006) also notes that a prime motivation for early corporate action is the belief that greenhouse gas regulation is imminent. But early corporate action is also motivated by the potential business impacts of regulation, including the need for new products and services to meet changing consumer and investor priorities. For example, Governor Schwarzenegger, quoted in a September 2006 press release, stated that:

> Some have challenged whether AB 32 is good for businesses. I say unquestionably it is good for businesses. Not only large, well-established businesses, but small businesses that will harness their entrepreneurial spirit to help us achieve our climate goals.
>
> (COG 2006a)

As a consequence of such messages, many business groups have accepted that early-acting states are more likely to attract investment and economic growth related to technological innovation in addressing climate change. While California is one of the largest US domestic producers of oil, gasoline and diesel, even this sector has not blocked the state from adopting ambitious climate policies. California's emissions policies are expected to reduce foreign imports of transportation fuel rather than fuel provided by in-state refineries. In addition, many aspects of California's economy would be hurt if action is not taken to reduce the magnitude of climate change, including real estate, insurance, recreation and agriculture, while professional and technical services, retail trade and finance are likely to benefit from investment in technologies to help reduce greenhouse gas emissions. Similarly, in New York, a summary of the state energy plan notes that:

> This Energy Plan positions New York to take advantage of technological developments among the most advanced uses of energy, and to participate in emerging markets for valuing and trading environmental attributes associated with energy use. In addition, implementation of this plan will stimulate job growth associated with the development of new technologies for the efficient production and use of a variety of energy resources and the expanded use of indigenous sources of power.
>
> (NYSERDA 2002)

In summary, the political and popular framing of climate change in key states as both an imminent, proximate threat and an economic opportunity (coupled with negative portrayals of the federal government response) has been pivotal in gathering broad-based political support for climate leadership. This task has been made somewhat easier in states like California and New York by the presence of persuasive communicators like Arnold Schwarzenegger in key state government positions; Schwarzenegger has even been prepared to use his celebrity status to challenge his own Republican Party's dominant discourses on climate change. Geographical allegiances have, thus, taken precedence over party political ones as advocacy coalitions in support of more stringent climate policies have gained the ascendancy in a growing number of US states.

Benefits, limits and strategies for state climate policies

As noted earlier, one of the key benefits of state leadership in environmental policy is that states can function as 'policy laboratories', developing initiatives that can serve as models for federal action. This is exactly what is happening with climate change through the various state policies and regional initiatives, where state programmes are already far ahead of federal policy. States are able to test out climate policies, see what works and what is problematic, and make recommendations for comprehensive legislation at the federal level. For example, in October 2007, agency heads from RGGI-participating states signed a letter to members of the US Congress, outlining recommendations for the design of a federal greenhouse gas cap-and-trade programme. They wrote:

> As states that have been actively engaged in the design and implementation of RGGI, we offer the enclosed design principles that we believe should be incorporated into a cap-and-trade program adopted as part of a comprehensive set of federal greenhouse gas emissions reductions policies... we encourage you to work with and learn from leadership states that are actively engaged in reducing greenhouse gas emissions.
>
> (RGGI 2007b)

Nevertheless, as Brewer and Pease discussed in Chapter 5, there are several generic political obstacles to reducing US greenhouse gas emissions, not least the economic and political influence of industry groups opposed to emissions reduction. In the states, the strength of political opposition relative to political support for greenhouse gas emissions reduction varies considerably. A challenge that is perhaps unique to California is the growing number of people in its southeastern deserts and central valley, areas that have hotter summer temperatures and higher energy consumption than the state's previously developed coastal population centres. California is seeking to counteract such effects by developing policies such as net-zero energy homes, energy efficiency, demand response, distributed generation and changes in land use planning, to help meet this challenge with low greenhouse gas design and technology. Bi-partisan support from political leaders, the building industry and the general public will be essential for the successful adoption and implementation of these policies (CEC 2007).

More generally, even in states with strong political support for action on climate change, the obstacles highlighted by Brewer and Pease in Chapter 5 could still undermine efforts to achieve the emissions cuts set by state and regional initiatives. State targets through to 2020 are also far below what will be required to prevent dangerous climate change. Attempts to introduce stronger policies in the future are likely to encounter stauncher opposition, although a lot will depend on future federal climate policy and the economic benefits of current policies being realized.

Given the magnitude of the changes needed and the political realities in the states, the political strategy generally adopted by state governments so far has been to start slowly, work out problems and bring in partners from neighbouring, like-minded states as a means of building political support. Another important tactic has been caution in the way emissions targets are presented in policy discourse and practice. Although long-term emissions reduction goals have been established in several states and regional initiatives, stronger commitments to deep greenhouse gas reductions at the state level would not have been politically feasible at this time and insistence on this might have stymied support for near-term reductions. This is especially the case in New York, where control of the state government is divided between the Democrats – who currently control the State Assembly and, as of 2007, the governorship – and the Republicans, who still hold a slim majority in the State Senate. Given the multi-decadal impacts of CO_2 emissions, leading states decided that it was important to begin reducing greenhouse gas emissions sooner rather than later, even if early targets remained relatively modest. Greater bi-partisan support and demonstration of the economic gains from investments in low-greenhouse gas technologies will be essential for deeper greenhouse gas reductions to become politically feasible, with the former depending strongly on the latter.

Another obstacle for state and regional policies is whether adequate staff, budget and resources will be available to implement and enforce emissions reduction policies. For example, New York's new Climate Change Office has only 12 employees, although this represents a significant increase from the one full-time member of staff employed in 2006. To help address this concern, more than 39 US states, seven Canadian provinces and two Mexican states have joined a voluntary mechanism to monitor greenhouse gas emissions reductions (similar to the CCAR) called The Climate Registry, a collaborative agency created in May 2007 to measure, track, verify and publicly report greenhouse gas reductions consistently and transparently (The Climate Registry 2008).

A further risk facing these state climate policies is the possibility that future federal administrations will develop federal policy with weak, pre-emptive targets and measures that inhibit the further development of state climate initiatives. It remains unclear at this time how the various state and regional programmes will fit into any future federal climate regime. The recent denial of California's request for a waiver under the Clean Air Act discussed above is just one example of the types of barriers an uncooperative federal government can impose on state efforts to reduce greenhouse gas emissions. Because of their growing investment in this policy field and the likely change in the federal position towards climate policy from 2009 onwards, some state governments are beginning to put pressure on the federal government to adopt a national market for carbon emissions reductions that is fully coordinated with state and regional policies. Governor Schwarzenegger highlighted this issue as part of the rationale for *Executive Order S-20-06* (COG 2006d):

> California, through the Western Governors' Association, has urged the President, Congress, the US Department of State, and other federal agencies to include the interests and expertise of the states as part of any national debate on the impacts of and efforts to reduce greenhouse gas emissions to ensure fully coordinated policies.

Reading between the lines, such tactics represent an attempt by state politicians to exploit a window of opportunity created by a federal policy vacuum on climate issues and an imminent change in administration to promote state involvement and state solutions in future climate policy discussions. If successful, this may have the effect of 'locking' the federal government into the states' agenda, creating a new trajectory for United States climate policy.

In addition to attempting to influence federal climate policy, some leading states are actively seeking coordinated approaches to emissions reduction on the international level. In October 2007, US and Canadian members of the WCI, Northeastern US members of the RGGI, and European Union member states and the European Commission signed the ground-breaking International Carbon Action Partnership (ICAP). New Zealand and Norway also joined ICAP on behalf of their emissions trading programmes (ICAP 2008). ICAP will help to advise governments and public authorities on how to adopt and manage cap-and-trade programmes. According to former New York Governor Eliot Spitzer at the signing in Lisbon:

> Global warming is the most significant environmental problem of our generation, and by establishing an international partnership, we are taking the vital steps to address this growing concern.
>
> (Goldberg 2007)

The advantage of this strategy, in addition to the general benefits of promoting a common approach to dealing with climate change, is that it helps to persuade apprehensive voters and businesses at home that their livelihoods and competitiveness will not be compromised by the decision to adopt more ambitious climate programmes.

This 'bottom-up strategy', where states take a leadership role in developing state, regional and even international climate initiatives, may be the only currently viable political strategy to enable emissions reductions until federal legislation is adopted and, perhaps, forced. States such as California and New York have already determined that it is ethically, scientifically and economically valid – and in their best interests – to act now on climate change, while the larger national consensus has yet to come to this conclusion. Leading states are, thus, beginning to move toward a low carbon economy at a pace and in a way that best reflects local, state and regional voter policy preferences. While these efforts are important in their own right, they are insufficient unless followed by additional measures, making it all the more important that short-term goals are met in ways that open the door to achieving the magnitude of reductions needed by 2050 to avoid dangerous climate change.

References

Baldassare, M., D. Bonner, J. Paluch and S. Petek (2007), *PPIC Statewide Survey: Californians and the Environment*, San Francisco: Public Policy Institute of California.

Barringer, F. (2008), 'California sues E.P.A. over denial of waiver', *The New York Times*, http://www.nytimes.com/2008/01/03/us/03suit.html?_r=1&oref=slogin&partner=rssnyt&emc=rss&pagewanted=print [27 January 2008].

Billings, L. (2007), *Climate Change: An Essay on Opportunity*, http://www.ncel.net/newsmanager/news_article.cgi?news_id=180 [19 January 2008].

Broder, J. and F. Barringer (2007), 'E.P.A. says 17 states can't set emission rules', *The New York Times*, http://www.nytimes.com/2007/12/20/washington/20epa.html [27 January 2008].

CACP (Clean Air-Cool Planet) (2005), *Governor Pataki receives 'Climate Champion' Award at 2005 Global Warming Solutions Conference*, http://www.cleanair-coolplanet.org/GlobalWarmingSolutions05/documents/Pataki.php [14 January 2008].

California Climate Change Portal (2004), *West Coast Governors' Global Warming Initiative*, http://www.climatechange.ca.gov/westcoast/index.html [14 January 2008].

CCAR (California Climate Action Registry) (2007), *About Us*, www.climateregistry.org [14 January 2008].

CEC (California Energy Commission) (1998), *Global Climate Change: Comments by William A. Keese, Chairman California Energy Commission*, http://www.climatechange.ca.gov/documents/98-01_keese_japan.html [14 January 2008].

CEC (2007), *Integrated Energy Policy Report*, http://www.energy.ca.gov/2007_energypolicy/ [10 March 2008].

CEPA (California Environmental Protection Agency) (2005), *Environmental Protection Indicators for California, 2004 Update*, http://www.oehha.ca.gov/multimedia/epic/pdf/EPICupdate2004a.pdf [14 January 2008].

COG (California Office of the Governor) (2006a), *Gov. Schwarzenegger Signs Landmark Legislation to Reduce Greenhouse Gas Emissions*, http://gov.ca.gov/index.php?/press-release/4111/ [14 January 2008].

COG (2006b), *Gov. Schwarzenegger, British Prime Minister Tony Blair Sign Historic Agreement to Collaborate on Climate Change, Clean Energy*, http://gov.ca.gov/index.php/press-release/2770/[14 January 2008].

COG (2006c), *Gov. Schwarzenegger, Premier of Manitoba, Canada Sign Agreement to Explore Linking of Future Carbon Trading Markets and Reduce Global Greenhouse Gas Emissions*, http://gov.ca.gov/index.php?/press-release/4920/ [14 January 2008].

COG (2006d), *Executive Order S-20-06*, http://gov.ca.gov/index.php?/executive-order/4484/ [15 January 2008].

COG (2007), *Governor Schwarzenegger Issues Statement on President Bush's State of the Union Address*, http://gov.ca.gov/index.php?/press-release/5227/ [14 January 2008].

DEC (New York State Department of Environmental Conservation) (2007), *New York Issues Draft Regulations for Cutting-Edge Initiative to Reduce Greenhouse Gas Emissions*, http://www.dec.ny.gov/environmentdec/39322.html [15 January 2008].

DEC (2008), *Energy and Climate: Regional Greenhouse Gas Initiative*, http://www.dec.ny.gov/energy/rggi.html [19 January 2008].

Goldberg, D. (2007), *Spitzer Joins International Climate Change Fight*, http://blog.syracuse.com/news/2007/10/spitzer_joins_international_cl.html [29 January 2008].

Hoffman, A. J. (2006), *Getting Ahead of the Curve: Corporate Strategies that Address Climate Change*, http://www.pewclimate.org/global-warming-in-depth/all_reports/corporate_strategies [10 March 2008].

ICAP (International Carbon Action Partnership) (2008), *Nations, States, Provinces Announce Carbon Markets Partnership to Reduce Global Warning*, http://www.icapcarbonaction.com/pr20071029.htm [20 January 2008].

ICLEI (International Council for Local Environmental Initiatives) (2008), *ICLEI – Local Governments for Sustainability*, http://www.iclei.org/ [20 January 2008].

IPCC (Intergovernmental Panel on Climate Change) (2007), *Climate Change 2007: Mitigation. Contribution of Working Group III to the Fourth Assessment Report of the Intergovernmental Panel on Climate Change*, IPCC: Cambridge: Cambridge University Press.

Iwanowicz, P. (2008), 'Manhattan to Lake Tear of the Clouds: Impacts on the Hudson Valley', *Paper to Climate Change: Science, Culture and the Regional Response Conference*, Fordham University, January 24–25, New York.

Klein, B. (2007), *Spitzer Planning New Climate Office*, http://www.legislativegazette.com/read_more.php?story=1975 [14 January 2008].

NCEL (National Caucus of Environmental Legislators) (2007), *Select State Actions to Address Climate Change*, http://www.ncel.net/newsmanager/news_article.cgi?news_id=184 [19 January 2008].

Northeast Climate Impacts Assessment (2006), *Climate Change in the US Northeast*, http://www.climatechoices.org/assets/documents/climatechoices/NECIA_climate_report_final.pdf [14 January 2008].

NYSERDA (New York State Energy Research and Development Authority) (2002), *Executive Summary, New York State Energy Plan*, http://www.nyserda.org/sep/sepexecsummary.pdf [29 January 2008].

NYSERDA (2008), *About NYSERDA*, http://www.nyserda.org [14 January 2008].

OCLI (Official California Legislative Information) (2002), *Assembly Bill No. 1493: Chapter 200*, http://www.leginfo.ca.gov/pub/01-02/bill/asm/ab_1451-1500/ab_1493_bill_20020722_chaptered.pdf [18 February 2007].

OCLI (2006), *Assembly Bill No. 32: Chapter 488*, http://www.leginfo.ca.gov/pub/05-06/bill/asm/ab_0001-0050/ab_32_bill_20060927_chaptered.pdf [25 February, 2007].

Office of the Governor, Wisconsin (2007), *Ten Midwestern Leaders Sign Greenhouse Gas Reduction Accord*, http://www.wisgov.state.wi.us/journal_media_detail.asp?prid=3027 [25 February 2008].

Pew Center (Pew Center on Global Climate Change) (2002), *States Act to Reduce Greenhouse Gas Emissions*, http://www.pewclimate.org/press_room/sub_press_room/2002_releases/pr_states_greenhouse.cfm [18 March 2008].

Pew Center (2007a), *What's Being Done... in the States: Latest News*, http://www.pewclimate.org/what_s_being_done/in_the_states/news.cfm [15 January 2008].

Pew Center (2007b), *Regional Initiatives*, http://www.pewclimate.org/what_s_being_done/in_the_states/regional_initiatives.cfm [17 January 2008].

Pew Center (2007c), *A Look at Emissions Targets*, http://www.pewclimate.org/what_s_being_done/targets [25 February 2008].

Pew Center (2007d), *Learning from State Action on Climate Change: December 2007 Update*, http://www.pewclimate.org/policy_center/policy_reports_and_analysis/state/ [19 January 2008].

Pew Center (2008), *State and Local Net Greenhouse Gas Emissions Reduction Programs*, http://www.pewclimate.org/what_s_being_done/in_the_states/database.cfm [19 January 2008].

Rabe, B. (2002), *Greenhouse and Statehouse: The Evolving State Government Role in Climate Change*, Pew Centre on Global Climate Change, http://www.pewclimate.org/press_room/sub_press_room/2002_releases/pr_states_greenhouse.cfm [15 January 2008].

Rabe, B. (2006), *Second Generation Climate Policies in the United States: Proliferation, Diffusion, and Regionalization, in Climate Change Politics in North America: The State of Play,* Occasional Paper Series, Canada Institute, Woodrow Wilson International Center for Scholars, http://www.wilsoncenter.org/topics/pubs/CI_OccPaper_ClimateChange.pdf [27 January 2008].

RGGI (Regional Greenhouse Gas Initiative) (2007a), *Overview of RGGI CO_2 Budget Trading Program,* http://www.rggi.org/press.htm [10 March 2008].

RGGI (2007b), *Agency Heads from RGGI Participating States Outline Recommendations for the Design of Federal Greenhouse Gas Cap-and-trade Program,* http://www.rggi.org/calendar.htm [10 March 2008].

RGGI (2008a), *About RGGI,* http://www.rggi.org/ [19 January 2008].

RGGI (2008b), *Model Rule,* http://www.rggi.org/ [19 January 2008].

RGGI (2008c), *Press Material,* http://www.rggi.org/ [19 January 2008].

Rothenburg, L. (2002), *Environmental Choices: Policy Responses to Green Demands,* Washington DC: CQ Press.

Schiller, S. (2007), 'Implications of defining and achieving California's 80 per cent greenhouse gas reduction goal', *Paper to Fourth Annual California Climate Change Conference,* Sacramento, California.

State of New York Executive Chamber (2005), *Directing State Agencies, and Authorities to Diversify Fuel and Heating Oil Supplies through the Use Of Biofuels in State Vehicles and Buildings,* http://www.ogs.state.ny.us/purchase/spg/pdfdocs/EO142_EEP.pdf [17 March 2008].

The Climate Registry (2008), *About Us,* http://www.theclimateregistry.org/index.html [24 February 2008].

United Nations Framework Convention on Climate Change (2007), *Total CO_2 Emissions, Excluding Removals from LULUCF in 2004,* http://unfccc.int/ghg_emissions_data/ghg_data_from_unfccc/time_series_annex_i/items/3843.php [19 January 2008].

US Census Bureau (2007), *Annual Estimates of the Population for the United States, Regions, States, and Puerto Rico,* http://www.census.gov/popest/states/NST-ann-est.html [14 January 2008].

US EPA (US Environmental Protection Agency) (2007a), *Energy CO_2 Emissions by State: State CO_2 Emissions from Fossil Fuel Combustion, 1990–2004,* http://epa.gov/climatechange/emissions/state_energyco2inv.html [15 January 2008].

US EPA (2007b), *The US Inventory of Greenhouse Gas Emissions and Sinks,* http://epa.gov/climatechange/emissions/downloads/2007GHGFastFacts.pdf [15 January 2008].

West Coast Governors' Global Warming Initiative (2004), *Staff Recommendations to the Governors, November,* http://www.ef.org/westcoastclimate/WCGGWI_Nov_04 per cent20Report.pdf [27 January 2008].

WCI (Western Climate Initiative) (2008a), *The Western Climate Initiative,* http://westernclimateinitiative.org/ [19 January 2008].

WCI (2008b), *The Western Climate Initiative: WCI Observers,* http://www.westernclimateinitiative.org/View_all_Observers.cfm [10 March 2008].

White House (2001), *President Bush Discusses Global Climate Change,* http://www.whitehouse.gov/news/releases/2001/06/20010611-2.html [15 January 2008].

Part III
Conclusions

14
Political Strategy and Climate Policy

Hugh Compston and Ian Bailey

Introduction

In the space of little more than two decades, climate change has been transformed from an obscure technical concern into an issue of major academic, political and public debate (Demeritt 2001: 307). As a result, reasonable degrees of confidence now exist on a number of key issues. Climate science has produced increasingly definitive statements about the existence, pace, extent and effects of human-induced climatic changes. The Stern Review, while contested in some quarters, provided strong indications that the economic costs of inaction or pure adaptation are likely greatly to outweigh those incurred by prompt and decisive action to cut and capture greenhouse gas emissions (Stern 2007). And various branches of the technical sciences and economics have developed innovative technologies and policy instruments to curb emissions. Despite these advances and sustained political attention, however, progress in cutting emissions remains disappointingly slow. The premise on which this book is based, and one that receives considerable support from the evidence reported in its constituent chapters, is that the main obstacles to more effective climate policies are essentially political: that, with a few exceptions, governments and other political authorities remain reluctant to take decisive action, even though most accept that strong measures are needed, because they fear that to do so would be politically damaging.

The aim of this book, accordingly, is to contribute towards resolving these political problems by identifying more clearly the main political obstacles to more vigorous action on climate change and the nature of political strategies that might enable these obstacles to be overcome or circumvented. We focus mainly on national (and, in federal countries, state-level) climate politics because this is where many substantive policy measures are formulated and implemented (Bailey 2007a), and because the international dimension of climate politics is already well-covered in the literature (Vogler 2005; Chasek *et al.* 2006).

In this chapter we draw on the findings of the preceding chapters in order to identify (i) the most prominent political obstacles to more effective action on climate change that have been encountered by the governments of the affluent countries that have contributed most to current greenhouse gas concentrations in the atmosphere; (ii) the most prominent political strategies used to date by these governments to try to bring climate change more under control; and (iii) the most promising political strategies for further action. Although we recognize that there can be no single, prescriptive political strategy to deal with the diverse situations facing affluent democracies, their experiences have enough in common for us to derive a number of general propositions about the political obstacles impeding climate policy and political strategies that may help to overcome or circumvent them.

Political obstacles

At least six major obstacles to implementing more radical climate policies can be identified: the perception that actions by individual countries make little difference to the progress of climate change; the continued influence of climate sceptics; a shortage of technically and economically efficacious solutions; the problem of competitiveness; fear of the electorate; and obstacles within government.

The perception that individual countries make little difference

It is widely accepted that, with the possible exceptions of the USA, the European Union (EU) as a whole and China, unilateral action by any one country to cut its greenhouse gas emissions, however radically, would not significantly slow global climate change. This underlines the importance of reaching a global or near-global agreement on strong policies. The Kyoto Protocol provided a first step, but the general view is that a much stronger international climate regime will be needed to deal effectively with the problem (Grubb *et al.* 1999; Helm 2005). Although the knowledge that unilateral action by any but the very largest countries will have little effect on climate change does not justify inaction at national level, since obtaining a global agreement that includes commitments by developing countries to limit emissions is likely to require substantial unilateral cuts by developed countries, the absence of a simple 'cause-and-effect' relationship between problem and solution within the domestic political setting that can be used to justify and legitimate strong climate policies is undeniably demotivating for national politicians. For this reason, lack of efficacy remains a serious obstacle to radical national action on climate change.

The influence of climate sceptics

In a large number of the countries examined, well-financed climate change sceptics continue to work to undermine support for climate policies by questioning the scientific consensus that climate change is being caused by human actions or by disputing the economic arguments favouring strong action. Where governments are sympathetic to such views, as was the case until recently in the USA, such groups can be a powerful brake on climate policy. Although the continued accretion of scientific evidence on climate trends and attribution appears to be progressively weakening the ability of climate contrarians to impede the strengthening of climate policies, except when they are in government (Chasek *et al.* 2006), their influence has been amplified and prolonged by influential media outlets which seek to enhance the news value of climate change by framing it as a debate in which the media has a duty to achieve 'balanced' coverage that gives equal exposure to supporters and opponents of the scientific consensus despite the overwhelming preponderance of evidence upholding the consensus scientific view (Boykoff 2007).

A shortage of technically and economically efficacious solutions

Despite much hype, a number of potential technological fixes, including hydrogen power, nuclear fusion, and carbon capture and storage, are from a technical point of view not yet ready to be implemented on a large enough scale to make a difference to climate change. In other cases the technologies are maturing but have yet to reach commercial viability, are inhibited by the lack of an agreed global carbon price or remain uncompetitive under existing pricing arrangements against (often subsidized) fossil-fuel technologies (this has been a particular problem for some renewable energies – see chapter by Michaelowa, this volume). Finally, certain technological or 'alternative' solutions remain highly contentious in terms of their contribution to greenhouse gas removals. Issues here include how to evaluate how much carbon forests really remove from the atmosphere in the short and long term, and whether, and if so the extent to which, schemes financed by Joint Implementation and the Clean Development Mechanism yield additional emissions savings. These factors combined limit the practical policy options open to governments.

The problem of competitiveness

Certain climate policy instruments, such as carbon/energy taxes, have the potential to increase production costs for affected firms, at least in the short term. To the extent that these firms export goods or compete with imports, and foreign firms are not subjected to the same costs, these policies can lead to a loss of international competitiveness that, if serious enough, would lead to insolvencies, cutbacks in investment and disinvestment, and thus to lower economic growth (or recession) and higher unemployment. Moreover, if affected industries relocate in significant numbers to countries without emissions constraints, overall industry emissions may not be reduced and may even increase (carbon leakage) (Barker and Ekins 2004).

Despite the lack of empirical evidence to support specific claims about competitiveness losses and carbon leakage arising from economic instruments (Ekins and Barker 2001), industry groups and associated concerns are rarely slow to bring these potential effects to the attention of politicians and officials. It is, therefore, not surprising that market-based instruments that are thought to erode competitiveness have often been avoided or diluted through the introduction of exemptions and concessions for energy-intensive or trade-exposed firms.

The obvious response is to level the playing field for domestic and foreign firms. While there is little immediate prospect of this happening at global level, the EU is a credible vehicle for addressing this issue for its member states, and in 2008 the Commission flirted with the concept of compensating for the increased costs incurred by European firms by imposing a carbon import tariff (Carbonpositive 2008). More ambitiously, the introduction of a common carbon tax across all EU countries would prevent competitive distortions within the Single European Market, although this would not help European-based companies vis-à-vis firms based outside the EU. The problem here is that some member states remain firmly opposed to ceding any further taxation powers to the EU. Fears about losses in competitiveness have also contributed to member state resistance to the tightening of emissions caps under the EU emissions trading scheme (EU ETS).

Fear of the electorate

Although the nature of representative democracy and the role of intermediary interest groups mean that there is no unmediated link between the preferences of domestic constituents and national climate politics, democratic governments cannot ignore electoral concerns when formulating domestic climate policies or their position on international negotiations (Sprinz and Weiß 2001). Growing public sympathy for the general notion of climate protection is arguably a major factor behind the emergence of greater cross-party agreement on the need for stronger climate policies in many countries. However, individuals tend to be less supportive of climate policies that directly or indirectly impose personal costs, or which impinge in other ways on personal freedoms, such as carbon/energy taxes, wind farms situated in their neighbourhood and lifestyle changes that are implied by climate policies such as restrictions on vehicle use. Employees whose jobs are perceived to be threatened by measures such as carbon taxes are also likely to object, and where these workers are unionized their objections may be taken up by trade unions. Whether or not climate policy is, or becomes, an inflammatory electoral issue, democratic governments which ignore these objections risk losing votes in the next general election that may make the difference between retaining and losing office. Moreover, where parties are polarized on particular climate policies, there is the further risk that an activist government would lose office to parties that have gained votes by promising to reverse these policies, so that even a noble sacrifice would be in vain. For these reasons, electoral opposition to certain climate policy measures remains an important constraint on the options available to national governments.

Obstacles within government

Although control of the executive in states such as the UK generally ensures safe passage through the parliamentary system for the government's preferred policies, in countries such as the USA legislatures can and do regularly block government proposals. Equally, in federal countries such as the USA, Canada and Australia, subnational governments possess constitutional powers that, depending on their specificity, enable them either to interfere with the implementation of the central government's preferred policies or to set independent agendas for climate policy. Tensions between state- and federal-level policies are particularly evident in Canada, the USA and Australia (see chapters by Brewer and Pease; Bailey and Maresh; Macdonald, this volume).

Another obstacle within government is the tendency of economic and energy ministries to oppose climate policies that are thought to have negative economic effects or threaten established interministerial relations. The ability of such ministries to block or dilute climate policies is enhanced by the fact that responsibility for critical areas such as energy and transport is generally located in economic rather than environmental ministries.

Of course the opposition of economic ministries is not enough to block strong climate policies if heads of government are determined to pursue them, but this is not always the case: lack of effective leadership at the top is often a further obstacle to greater progress on climate change. One example is Chancellor Kohl blocking the introduction of an ecotax in 1993; another is the drift at the top described in the chapter on Canada.

Current political strategies

The main political strategies currently being used by the governments of the countries reviewed in this book to strengthen climate policy can be divided seven broad categories: (i) efforts to reach global agreement on policies to control climate change; (ii) reports and targets; (iii) climate policies on which all major relevant political actors can agree; (iv) incremental policy changes; (v) taking advantage of weather-related natural disasters; (vi) framing climate policies in terms of other desired policy objectives; and (vii) in terms of policy instruments, a focus on information provision, technological fixes, renewable energy, energy efficiency, voluntary agreements, and, in some countries, emissions trading and carbon/energy taxes.

Before looking at these, it is important to make the point that one major advantage for advocates of more decisive action on climate change is the broad level of support that exists in most affluent democracies among political and economic elites for action to limit the effects of climate change. Whether this support is motivated by environmental, social or pecuniary concerns, such a broad level of agreement on objectives is not characteristic of all policy areas (for example, taxation or employment policy).

Efforts to reach global agreement on climate change

Given the need to develop strong global agreements before dangerous human interference with the climate system occurs, efforts in the international arena need to be mentioned even though international climate politics is not the primary focus of this book. Climate scientists are clear, and most politicians concede, that the political targets set in the Kyoto Protocol, even if achieved, are nowhere near sufficient to bring climate change under control.

One aspect of this effort has been efforts at least partially to enact the principle of common but differentiated responsibility enshrined in the United Nations Framework Convention on Climate Change, with most developed countries committing themselves as Annex B signatories to the Kyoto Protocol to reducing their greenhouse gas emissions in advance of action by developing countries in recognition of the fact that they have used vast amounts of fossil fuels to drive economic expansion and, thus, have contributed most to current anthropogenic greenhouse gas concentrations. Coupled with this is the argument that these countries possess the legal, political, economic and technological capability to lead the mitigation effort, lessons from which could also be used to assist developing countries to advance economically and socially without following similar emissions trajectories. Although the reality of international climate politics is often rather less noble, with governments tussling to limit their climate liabilities and orient international policy towards their domestic economic interests (see chapters by Davenport; Bailey and Maresh), genuine efforts to reach more far-reaching global agreements continue to be undertaken.

Reports and targets

Although government reporting and target setting can in some cases be used as a substitute for action, when not used cynically reports and targets can play an important role in diagnosing the scale of the problem, specifying required outcomes, and in so doing provide a statement of intent, build support for action and inform the structure and design of policy instruments. Cases in point include the EU's commitment to reduce its aggregate emissions to 20 per cent below 1990 levels by 2020, and by 30 per cent if other major emitters do likewise, and the target to reduce emissions by 60 per cent by 2050 contained in the UK Climate Change Bill, with a further review of arguments to strengthen this to 80 per cent (see chapters by Damro and MacKenzie; Lorenzoni *et al.*, this volume). That said, there is no guarantee that targets will be met, and one recurrent finding emerging from the chapters in this book is that they are often missed.

Focus on climate policies on which all major relevant political actors can agree

One way that governments can reduce the political damage associated with climate policies is by obtaining the prior agreement of the main affected political actors and, implicitly, of the electorate as indicated by the results of opinion polls. This approach has pervaded the climate politics of many of the case-study countries. One indication of this is the ubiquity of voluntary agreements, whereby industry groups undertake to reduce their emissions in exchange for the non-imposition or delay of legal requirements or economic instruments. Such strategies can be especially important in countries in which relevant legal powers are divided between the executive and the legislature and/or between national and subnational levels of government. Although examples exist of climate policies being imposed by governments over industry and political opposition, such as the German ecotax in the late 1990s, in general this has occurred only after lengthy debates and/or periods of reliance on voluntary commitments, and the imposition of such policies have often been accompanied by concessionary measures.

The extent to which consensus strategies are capable of producing effective action – that is, emissions reductions significantly above business-as-usual – remains open to debate. In addition, if we assume that affected actors generally prefer less to more action unless 'no regrets' outcomes can be clearly demonstrated, it follows that the perceived need to obtain broad agreement will generally result in weaker climate policies than if governments that possess the necessary legal powers override opposition, and that radical policies which impose significant costs on one or more key actor groups or significant sections of the electorate will be ruled out of consideration.

Although the use of consensus strategies has varied considerably between countries and types of climate policy (Thalmann and Baranzini 2004), the experience of affluent democracies demonstrates that a wide range of climate policies exists on which broad agreement can be reached, depending on circumstances and how policy mixes are designed and negotiated. Consensus strategies are especially important in the early stages of climate policy as a means of binding key actors (especially industry groups) to the principle of emissions reduction while avoiding excessive political costs. On the other hand, reliance on consensus decision-making can create path dependencies in the way climate policies are negotiated that are difficult to break and can stall progress on stronger measures: once the policies on which agreement can be reached have been implemented, persistence with a consensus strategy can impede further progress due to the effective veto that the perceived need for agreement gives to relevant stakeholders.

Incrementalism on many fronts

One widely shared belief among policy analysts and practitioners is that incremental policy changes usually elicit less political opposition than radical policy changes, and that incremental changes can create a platform for more major policy changes (Oliver and Pemberton 2004). Another important argument for small-scale reforms across a wide range of areas is that it reduces the chances of major policy failures and allows policy innovations to be tested before being disseminated more broadly. This is especially pertinent to climate policy; as Bryner and Barker both point out in earlier chapters, there is sufficient uncertainty about the timing, severity and costs of climate change, and about the efficacy of different policy responses, to justify exploring all perceived options rather than committing early and heavily to any individual one. It is, therefore, not surprising that most governments in affluent democracies have tended to implement a wide variety of relatively weak policies rather than adopting a few big radical measures. Germany is particularly renowned for its long lists of climate measures. On the other hand, emissions trading schemes introduced in the EU, some US states and Australia do constitute major departures from past practices (see chapters by Damro and MacKenzie; Chatrchyan and Doughman).

Taking advantage of weather-related natural disasters

Even though it is impossible to be certain that individual weather-related disasters are a direct result of climate change, there is evidence that scientific research linking extreme weather events to climate change has fuelled media speculation on the subject when such events occur (Boykoff 2007). The resulting spikes in public concern about climate change then create windows of opportunity for governments to introduce or strengthen climate policies without sustaining as much political damage as might be the case at other times. In some cases governments might even benefit politically. At the same time, it is important to note that these windows of opportunity are only temporary and tend to close as the event becomes more distant in time and the media move on to other issues (Lorenzoni *et al.*, this volume). Nevertheless, there is evidence that in some cases these opportunities have been grasped by governments, as happened in Germany in 2002 when the Elbe floods were instrumental in the government introducing new targets, and in Sweden after the Gudrun storm (see chapters by Michaelowa and Friberg).

Framing climate policy in terms of other policy objectives

An increasingly common device used by governments to broaden support for particular climate policies is to stress their contribution to the achievement of other social and economic objectives (so-called ancillary and co-benefits). Expansion of energy generation from renewable sources, for example, contributes to energy security and employment, while measures to encourage people to switch from private cars to public transport would be expected to reduce traffic congestion as well as reducing emissions. The advantage of this approach is that actors who support these other objectives can be recruited to swell advocacy coalitions favouring, directly or indirectly, the objectives of climate policy. This can be especially important in relation to unpopular climate policies and in countries where climate scepticism is strong. In extreme cases, climate objectives may have to remain in the background in order to make measures more palatable to key actors or electorates.

Instrument choice and design

Although governments in affluent democracies have used a diverse range of policy instruments to control greenhouse gas emissions, several instruments are used in most or all of the countries examined in this book and some are also deployed at EU level. In general, governments have gradually moved from relying on voluntary agreements towards the use of economic instruments underpinned by targets and legal standards.

First, governments in all these countries use information dissemination and awareness-raising techniques, such as energy-efficiency ratings, labels, auditing and advice, to try to persuade firms and households to reduce emissions by informing them of the greenhouse gas implications of their actions and how changes in behaviour could reduce their carbon footprint.

Second, all the governments surveyed use subsidies and grants for research and development to encourage the design and dissemination of new technologies to limit or capture emissions. Examples of these include hydrogen technologies, nuclear fusion, and carbon capture and storage.

commercialization: 'the technologies upon which any emissions reduction strategy depends simply will not be available at a competitive cost at the time when they could make a significant difference'. A massive expansion of nuclear power is one option, though by no means the only possible *grand projet*, although official support for this appears to be growing in the UK and affecting political discourse in other countries like the Netherlands. The big advantage of nuclear power over renewables is that nuclear power stations can generate large amounts of reliable baseload energy whereas the output of electricity generation from renewable sources, such as the sun and wind, is more intermittent. In the absence of adequate storage mechanisms, using renewable energy to supply more than a certain percentage of electricity creates a risk of electricity shortages during periods of peak demand if weather conditions are unfavourable. The big problem for nuclear power, however, is the strength of public opposition, especially in countries such as Germany, Sweden, Australia and Greece. A big expansion of nuclear power would also necessitate overcoming serious economic and technical problems in relation to, among other things, the long-term storage of radioactive waste. On the other hand, the experience of France during the 1970s and 1980s demonstrates that, in certain (historical and perhaps not to be repeated) circumstances, nuclear power can be massively expanded without obvious negative political repercussions.

Fourth, in countries such as the UK, an extension of the tradable permits methodology to the individual level in the form of *personal carbon allowances* (PCAs) has been mooted (see Lorenzoni *et al.*, this volume). Under such schemes, individuals would be allocated PCAs for specified energy purchases, with those who consume less than their allocation selling their surplus allowances to high carbon consumers. Although doubts have been raised about the complexity, controllability, monitoring, public acceptability and possible manipulation of PCAs, they appear to provide a fair and equitable means of giving financial incentives for individuals and communities to reduce their energy use and carbon footprints. Lorenzoni and her colleagues accordingly suggest that governments should embark immediately on community experiments with PCA schemes so that their practical consequences can be explored in advance of any large-scale implementation.

Fifth, EU member states could move to counter any competitive disadvantages suffered by EU-based firms as a consequence of the costs of national climate policies by moving to create a level trade playing field vis-à-vis countries with less stringent climate policies by introducing *carbon import tariffs*. This could also create incentives for countries that have not ratified the Kyoto Protocol to do so, and for those that do not have binding commitments under it to implement stronger climate policies. The European Commission's initial proposal for carbon import tariffs was opposed by its own Trade Commissioner, Peter Mandelson, because of the complications it would pose for international trade negotiations, and a decision has been reserved until the next EU ETS review in 2011 (Carbonpositive 2008). However similar proposals are beginning to be mooted in other countries (see, for example, Courchene and Allan (2008) in relation to Canada).

Finally, affluent countries could *make financial aid for developing countries conditional on climate policy commitments* on their part, with the United Nations acting as an honest arbiter of the process (Sandbrook 1997), although such 'climate-for-development' conditionality would be bitterly opposed by many developing nations.

New political strategies

It is clear, then, that there are a number of new policy options that activist governments could adopt. But will such approaches be acceptable to key actor groups and electorates, or will resistance to their attempted implementation inflict serious political damage on governments, or even loss of office followed by the reversal of these policies by replacement governments? Democratic governments cannot be expected to act unless they can answer these questions to their own satisfaction. It is for this reason that it is insufficient to identify technically and economically viable policy options: overtly political strategies to facilitate their introduction are also required. This section outlines three new strategies that the findings of the country chapters, plus insights from elsewhere, suggest may enable governments to introduce more radical policies without cutting their own throats: reform of economic and environmental governance; use of a spillover strategy; and the selective imposition of climate policies.

Governance reform

There are a number of adjustments to the structure and activities of governance that governments can employ to promote more effective action on climate change.

An obvious first priority is to take steps to improve the credibility and legitimacy of climate policies *by improving the processes used to measure emissions and set, monitor, and report on emissions targets* for different sectors of the economy and for countries as a whole. Among other things this would involve (i) developing and disseminating more accurate methods for measuring emissions; (ii) more vigorous efforts to reach agreement on common accounting rules for comparing and tracking countries' emissions over time; (iii) finding ways to counter manipulation of these rules by governments and industry groups that wish to reduce the stringency of requirements; (iv) improving transparency and visibility in emissions reporting by means such as making public disclosure of emissions data and targets mandatory for major companies and sectors (Walton 2000); and (v) strengthening links between emissions monitoring and target setting, a point highlighted in the House of Commons Environmental Audit Committee's (2008) report into the UK Climate Change Levy and Agreements, where it was noted that Climate Change Agreements had not been tightened despite the fact that the first set of performance results in 2004 had revealed that they were too weak.

A second priority is clearer identification of the exact policies needed if countries are to achieve the magnitude of emissions reductions needed to bring climate change under control. The approach of most governments thus far has been to work forward by setting relatively modest short-term targets and more ambitious long-term ones without developing a clear view of the policies required to achieve the latter. One way of remedying this deficiency would be through *more systematic envisioning of the key characteristics of a low-carbon society and economy* and the use of scenarios and backcasting techniques to identify the key policy steps and timings required. The use of scenarios and backcasting is discussed more fully by Berkhout *et al.* (2002), who argue that scenario approaches provide a useful way of engaging political and non-state actors in the process of social and organizational learning that sets a frame for iterative self-reflection, change and adaptation.

The third priority is to address tensions between environmental and economic ministries on climate policies. Several options exist here. The first would be to *integrate economic and climate governance*, for example by moving energy and transport into an environmental ministry, although this could lead to climate policy being stifled within the ministry even before issues reach interministerial negotiations. An alternative would be to *create a separate climate ministry*, as has occurred in Australia, to elevate the standing of climate issues in cabinet discussions, although this still provides no guarantee of overcoming interministerial insularities and power disparities. A third possibility – but with similar caveats – would be to *create ministerial climate policy steering groups* to promote a more 'joined up' view of relations between climate change and other policy areas, as already exist in Finland, New Zealand and the UK (Begg and Gray 2004). A further device for integrating the climate consequences of government decisions into policy-making would be to *add the minimization of greenhouse gas emissions to existing decision criteria* in all areas of public policy, and to make the greenhouse gas implications of all policy options explicit by *including carbon costs in all government policy and investment decisions*. This aligns particularly well with the reasoning underpinning the five-year carbon budgets proposed in the UK Climate Change Bill.

Another way of altering the institutional dynamics of climate decision-making would be to raise the profile of environmental NGO and academic representatives in decision-making. In several of the countries examined, climate policy-making remains relatively closed and/or provides better opportunity structures for industry groups to influence policy than exist for environmental NGOs. For example, industry groups often have greater representation on key consultative committees, whereas NGO and, to a lesser extent, academic input is often restricted to consultation documents and public hearings. Industry groups also have more money to pay for lobbying of decision makers. One way to balance this disproportionate industry influence in policy-making would be to *require that independent experts and environmental NGO representatives be included on all relevant committees on which industry representatives sit*.

The country chapters also provided strong evidence of the importance of having committed individuals in key positions, in the words of Zito (2000: 11) to: 'manipulate the constellation of ideas and interests and take advantage of favourable institutional structures' to promote more ambitious climate policies. Examples of the significant role that individuals can play include Klaus Töpfer's support for ecotaxation as German Environment Minister during the 1990s, Arnold Schwarzenegger's role in driving forward the climate agenda at state, national and international levels as Governor of California, and, in a different guise, Tony Blair's inclusion of climate change as a priority at the 31st G8 Summit in Gleneagles in Scotland. For voters, this means voting for relatively green parties and leaders; for heads of government it means *placing able and committed colleagues in key ministerial posts*.

A fourth strategy that has been used with some success in countries such as Germany and Sweden, and at state level in the USA, is to *maximize the transparency of climate policy initiatives*, that is, to bolster public support for policies that governments feel could be made popular by giving clear explanations about their intended outcomes and the stages involved in the policy-making process. This statement may seem self-evident; however, lack of transparency in EU climate policy has frequently been blamed for conflicts with the member states, public antipathy or ignorance towards climate initiatives, and even for increasing the influence of special interest groups over decision-making (Michaelowa 1998; Sprinz and Weiβ 2001). During Phase two of the EU ETS, for instance, several member states accused the European Commission of failing to disclose the methodologies used to justify reductions in national emissions caps (see chapter by Damro and Mackenzie). Although transparency may be less relevant for policies that are intrinsically unpopular, like ecotaxes, there is some evidence that clear communication about the aims and long-term direction of tax measures can help to reduce opposition here too. The introduction of the UK Climate Change Levy in 2001, for instance, was preceded by lengthy consultations with industry and other interest groups on the design of the tax. Although the government was criticized for granting too many concessions, the process did enable industry groups to air grievances and to become accustomed to the tax, leading arguably to a less hostile reaction than if the policy-making process had been less transparent (Bailey 2007b).

A fifth and final tactic in relation to governance reforms is to *distribute any costs imposed by climate policies as equitably as possible.* If people can see that these costs are distributed fairly, so the argument goes, they will be more likely to accept them, while granting generous concessions to powerful groups – a common device to reduce opposition to unpopular measures like taxes – has the capacity to tarnish the reputation of climate policies even after these have been introduced. The German ecotax, for example, is unpopular at least in part because a majority of survey respondents consider it to be socially unfair because the majority of the tax burden fell on households and transport after industry gained various concessions (Weidner 2008: 20). This perception has persisted despite the reduction of these concessions once the government deemed that industry had been given enough time to adjust to the tax. This suggests that close attention to different groups' ability to pay, their direct responsibility for emissions and the balance between 'essential' and 'luxury' emissions (along with clear communication of how these issues will be managed), may play a valuable role in minimizing antagonism to measures such as taxation. At the same time, the extent to which equity strategies can be used in practice is likely to be limited due to the fact that distributing costs fairly is often incompatible with consensus strategies that involve giving powerful actors concessions in order to obtain their agreement.

Spillover strategies

The concept of policy spillover has been widely employed in the political sciences to describe functional or political pressures to strengthen existing policies or introduce new policies. Two related but distinct senses of the term are relevant here.

First, there is the idea of international spillover whereby the introduction of climate policies in one country leads via means such as technological diffusion to similar policies being enacted in other countries (Grubb *et al.* 2002). Policy spillover in this sense is very similar to the idea of policy transfer (Dolowitz and Marsh 1996). Among other things this implies that *governments should choose policies that are relatively likely to transfer to other countries,* for example policies that encourage renewables technologies that can be easily adapted and utilized in developing countries.

The second sense of spillover is that used in European integration theory, in which one step in integration, once implemented, increases functional and/or political pressure on decision makers to take the next step. Thus, according to this logic, the establishment of the Customs Union increased pressure to establish a Single Market, which in turn increased pressure to establish a single European currency (McCormick 2001).

Applied to climate policy, spillover in this sense refers to policies that, once established, would obviously work better if strengthened or joined by further climate policies, or which expand the range of political actors who want further strengthening of climate policy. It is worth noting here that spillover pressure in this sense already exists in that the existence of the Single Market and monetary union creates pressure for EU-wide climate policies to avoid market distortions. It is important to note, however, that there is no automaticity to spillover: although the logic of spillover points to an EU-wide carbon/energy tax, for example, such a tax has not (yet) eventuated.

Perhaps the minimum case of spillover in this sense is where introduction of a policy creates pressure to continue with this policy. To some extent inertia, the relative ease of staying with the status quo and the benefits experienced from existing policies means that this effect is often present up to a point once a policy is introduced, but what we have in mind here are policies that create more substantial pressure for their own continuation, that is, policies that once established are very difficult to reverse, such as the signing of long-term contracts that commit countries to major investments in the restructuring of energy supply (whether renewables, nuclear power or a combination of the two).

An example of a policy with potential for positive spillover would be one that leads to increased investment and employment in the renewables sector, as these employment effects might be expected to increase political support for further expansion of this sector and in this way increase pressure on governments to take further steps to facilitate this (Markandya and Rübbelke 2003).

The aim of a deliberate spillover strategy, then, would be to select and implement policies that (i) increase the pressure on other governments to implement similar policies and (ii) increase pressure for stronger such policies to be implemented at home.

Selective imposition of more radical policies

We have already seen that focusing on measures on which the agreement of powerful actors can be obtained is an effective way of strengthening climate policy up to a point. Once relatively uncontroversial policies have been negotiated and implemented, however, continued acceptance of the need to obtain broad agreement impedes the introduction of more radical measures by giving all stakeholders an effective veto on government action. This, combined with the fact that consensus strategies have not (yet) delivered the levels of emissions cuts needed for effective mitigation, suggests that as time goes on consensus-based policies are likely to suffer diminishing returns unless they result in technological innovations that yield major no- or low-cost emissions cuts, or natural disasters change the minds of hitherto recalcitrant actors.

It seems inevitable, therefore, that at some point governments will have to introduce at least some more radical policies against the wishes of powerful actors and/or voters – in other words, governments will need to adopt a strategy of selective imposition of more radical policies.

Such a strategy is, of course, much riskier than waiting for agreement among all major players. Although for this reason it may seem out of place in a section on strategies that may enable governments to take stronger action while *avoiding* serious political damage, it has to be listed because (a) without such a strategy the prospects of controlling climate change appear to be bleak and (b) there may well be ways in which governments can implement such a strategy without suffering too much damage. It is also important to note that we have already seen cases of governments imposing climate policies over vociferous opposition without apparently being punished to any significant extent, two examples being the German ecotax of 1999 and the London congestion charge.

The question is therefore one of developing political tactics to manage this risk. Space does not permit a full account of what these might be, but we would like to mention just a few.

One well-known tactic is to *impose unpopular measures during the early years of an administration* in order to allow time for opposition and critical media coverage to subside and for the benefits and real (as opposed to inflated projected) costs to become apparent.

A second approach is to *target economic sectors that can pass on extra costs to consumers*. In its ideal form, this tactic can allow for the gradual internalization of environmental costs into the price of energy, goods and services without government being blamed directly. One example of this is measures targeting the energy generation and supply sectors, which tend not to experience strong international competition and where, in deregulated markets, providers will also compete with each other on price, thereby theoretically avoiding profiteering. It is important to note, however, that this does not always work, as the media is often quick to publicize how carbon/energy taxes, for example, lead to higher prices for consumers.

A third tactic would involve governments introducing just a few major changes in a small number of selected, high-impact areas so as to keep most groups onside by *concentrating losses on just a few*, in particular those groups that are able to inflict the least political damage via the ballot box or are least likely, despite rhetoric to the contrary, to reduce or withdraw investment.

A fourth tactic would be to *compensate key players* for their losses. This option featured strongly in Douglas Macdonald's review of policy options for Canada, where he suggests that oil-rich Alberta will need to be compensated in order to persuade it to set aside its objections to more progressive climate policies. This is, of course, part and parcel of consensus-seeking strategies, but it can also be used in imposition strategies to weaken resistance: those affected may still oppose new measures, but compensation is likely to reduce their willingness and ability to mobilize resistance. The main problem with this approach is that any reduction in political damage due to providing compensation for powerful actors, in particular those that have become affluent by emitting large quantities of greenhouse gases, could be offset by damage incurred as a result of being seen to implement inequitable policies. Recognizing these difficulties, Macdonald adds that compensation measures must be accompanied by a clear 'tough-love message' to sub-national administrations and corporations that it understands their position and will assist them to adjust but will no longer tolerate inaction or obstructive tactics. Provided compensation packages are designed as transitional measures only, therefore, they do appear to provide a promising way for governments to weaken the resistance of key political opponents.

Concluding remarks

We would like to begin our concluding remarks by again paying tribute to the contributors to this volume for providing informed and insightful analyses of the political obstacles to more progressive climate policies in affluent democracies and of strategies that governments might employ to address these without sustaining significant political damage. It is on the basis of these that we have been able to put together our account of political strategies to enable governments to implement more effective climate policies. Although this account is by no means definitive – academic analysis in this area is surprisingly underdeveloped, and in any case no unifying set of prescriptions could cope with the diverse socio-economic and political situations in the countries studied – the strategies drawn from lived political experiences in affluent democracies that we have identified merit serious consideration by proponents of stronger action on climate change.

What is certain is that the 'old' politics of climate policy has a limited shelf life. Existing political strategies have initiated a process and made some progress in reducing emissions, but not on the scale required to bring the problem of climate change under control. It may be that some combination of scientific research, economic instruments and new technologies will provide a robust solution to climate change. However, putting these into practice will still require politicians at local, national and international levels to confront, and find innovative strategies to deal with, the political obstacles that have so far hindered progress towards effective climate governance. We leave you with a summary of these:

- *Refinement of current strategies*, in particular further efforts to reach global agreement; better reporting of climate change; clearer communication of the policy instruments that are needed; stricter emission and policy targets; identification and introduction of more policies on which all powerful actors can agree; incremental strengthening of existing policies; preparation of measures that can be implemented swiftly in response to public concern following extreme weather events; continued emphasis on the contribution of climate policies to other policy objectives such as energy security; and more vigorous use of existing policy instruments, especially economic instruments and financial incentives to promote technological innovations and renewable energy production, more stringent voluntary agreements, and extension of emissions trading;

- *Exploration of new policies*, in particular more stringent energy-efficiency regulation and increased financial incentives for energy-efficiency improvement; *grand projet* style state investment in new infrastructure; personal carbon allowances; and carbon import tariffs;
- *Governance reform*, in particular improving measurement of emissions; more systematic envisioning of what a low carbon society would look like; integrating economic and environmental governance; providing seats for independent experts and environmental NGOs on all committees on which industry is represented; placing able and committed individuals in key posts; improving the transparency of potentially popular initiatives; and distributing costs more equitably;
- *Spillover policies*: choosing policies that are easily transferable to other countries, difficult to reverse once introduced, or create pressure for their own strengthening or the introduction of related measures;
- *Selective imposition of more radical policies*, taking steps to minimize political damage by means such as introducing strong policies early in each electoral term, targeting economic sectors that can pass on extra costs to consumers, targeting losses on small sections of society and compensating powerful actors.

References

Asheim, G., C. Bretteville Froyn, J. Hovi and F. Menz (2006), 'Regional versus global cooperation for climate control', *Journal of Environmental Economics and Management* 51, 93–109.

Bailey, I. (2007a), 'Climate policy implementation: geographical perspectives', *Area* 39, 415–17.

Bailey, I. (2007b), 'Market environmentalism, new environmental policy instruments, and climate policy in the United Kingdom and Germany', *Annals of the Association of American Geographers* 97, 530–50.

Barker, T. and P. Ekins (2004), 'The costs of Kyoto for the US economy', *The Energy Journal* 25 (3), 53–71.

Begg, D. and D. Gray (2004), 'Transport policy and vehicle emission objectives in the UK: is the marriage between transport and environment policy over?' *Environmental Science & Policy* 7, 155–63.

Berkhout, F., J. Hertin and A. Jordan (2002), 'Socio-economic futures in climate change impact assessment: using scenarios as "learning machines"', *Global Environmental Change* 12, 83–95.

Boykoff, M. (2007), 'Flogging a dead norm? Newspaper coverage of anthropogenic climate change in the United States and the United Kingdom from 2003 to 2006', *Area* 39, 470–81.

Carbonpositive (2008), *EU Flirts with Carbon Import Tariff*, http://www.carbonpositive.net/viewarticle.aspx?articleID=952 [1 May 2008].

Carraro, C. (2000), 'The economics of coalition formation', in Gupta, J. and M. Grubb (eds), *Climate Change and European Leadership: A Sustainable Role for Europe?* Guildford: Springer, pp. 135–56.

Chasek, P., D. Downie and J. Welsh Brown (2006), *Global Environmental Politics*, Boulder CO: Westview.

Courchene T. and J. Allan (2008), 'Climate change: the case for a carbon tariff/tax', *Policy Options* March 2008, 59–64.

Demeritt, D. (2001), 'The construction of global warming and the politics of science', *Annals of the Association of American Geographers* 91, 307–37.

Department for Communities and Local Government (2008), *The Code for Sustainable Homes: Setting the Standard in Sustainability for New Homes*, London: Communities and Local Government.

Dolowitz, D. and D. Marsh (1996), 'Who learns from whom: a review of the policy transfer literature', *Political Studies* 44, 343–57.

Ekins, P. and T. Barker (2001), 'Carbon taxes and carbon emissions trading', *Journal of Economic Surveys* 15, 326–75.

Grubb, M., C. Vrolijk and D. Brack (1999), *The Kyoto Protocol: A Guide and Assessment*, London: The Royal Institute of International Affairs.

Grubb, M., C. Hope and R. Fouquet (2002), 'Climatic implications of the Kyoto Protocol: the contribution of international spillover', *Climatic Change* 54, 11–28.

Helm, D. (ed.) (2005), *Climate Change Policy*, Oxford: Oxford University Press.

House of Commons Environmental Audit Committee (2008), *Reducing Carbon Emissions from UK Business: the Role of the Climate Change Levy and Agreements*, London: The Stationery Office.

Lorenzoni, I., N. Pidgeon and R. O'Connor (2005), 'Dangerous climate change: the role for risk research', *Risk Analysis* 25, 1387–98.

Markandya, A. and D. Rübbelke (2003), 'Ancillary benefits of climate policy', *Fondazione Eni Enrico Mattei (FEEM) Working Paper* No. 105.2003.

McCormick, J. (2001), *Environmental Policy in the European Union*, Basingstoke: Palgrave.

Michaelowa, A. (1998), 'Impact of interest groups on EU climate policy', *European Environment* 8, 152–60.

Oliver, M. and H. Pemberton (2004), 'Learning and change in 20th-century British economic policy', *Governance* 17, 415–41.

Prévot, H. (2007), *Trop de pétrole: énergie fossile et réchauffement climatique,* Paris: Seuil.

Rayner, S. (2004), *Memorandum submitted to the House of Commons Environmental Audit Committee*, Session 2004–05, 15 December 2004, http://www.publications.parliament.uk/pa/cm200405/cmselect/cmenvaud/105/4121506.htm [Accessed 24 July 2007].

Sandbrook, R. (1997), 'UNGASS has run out of steam', *International Affairs*, 73, 641–654.

Sprinz, D. and M. Weiβ (2001), 'Domestic politics and global climate policy', in Luterbacher, U. and D. Sprinz (eds), *International Relations and Global Climate Change*, Cambridge MA: MIT Press, pp. 67–94.
Stern, N. (2007), *The Economics of Climate Change: The Stern Review*, Cambridge: Cambridge University Press.
Thalmann, P. and A. Baranzini (2004), 'An overview of the economics of voluntary approaches in climate policies', in Baranzini, A. and P. Thalmann (eds), *Voluntary approaches in Climate Policy*, Cheltenham: Edward Elgar, pp. 1–30.
Vogler, J. (2005), 'The European contribution to global environmental governance', *International Affairs* 81, 835–50.
Walton, J. (2000), 'Should monitoring be compulsory within voluntary environmental agreements?' *Sustainable Development* 8, 146–54.
Weidner, H. (2008), 'Climate change policy Germany: capacities and driving forces', paper prepared for the Workshop 'The Politics of Climate Change', *ECPR Joint Sessions, Rennes*, 11–16 April.
Zito, A. (2000), *Creating Environmental Policy in the European Union*, Basingstoke: Macmillan.

Index

CPI Antony Rowe
Chippenham, UK
2017-03-28 21:28